Aluminum-Silicon Casting Alloys

Atlas of Microfractographs

Małgorzata Warmuzek

ASM International®
The Materials Information Society

Materials Park, OH 44073
www.asminternational.org

First printing, May 2004

Comments, criticisms, and suggestions are invited and should be forwarded to ASM International.

Prepared under the direction of the ASM International Technical Books Committee (2003-2004), Yip-Wah Chung, Chair

ASM International staff who worked on this project include Charles Moosbrugger, Acquisitions Editor; Bonnie Sanders, Manager of Production; Nancy Hrivnak and Jill Kinson, Production Editors; and Scott Henry, Assistant Director of Reference Publications.

Library of Congress Cataloging-in-Publication Data

Warmuzek, Małgorzata.
Aluminum-silicon casting alloys: an atlas of microfractographs / Małgorzata Warmuzek.
p. cm
Includes bibliographical references and index.
ISBN 0-87170-794-2
1. Aluminum alloys—Fracture—Atlases. 2. Aluminum alloys—Metallography—Atlases. I. Title.

TN693.A5W37 2004
669′.722—dc22

2004040998

ISBN: 0-87170-794-2
SAN: 204-7586

ASM International®
Materials Park, OH 44073-0002
www.asminternational.org

Printed in the United States of America

Contents

About the Author

Małgorzata Warmuzek earned her master of science degree at the Academy of Mining and Metallurgy, Kraków, Poland, in physical metallurgy and heat treatment in 1974. She earned her doctorate from the Foundry Research Institute, Kraków, in physical metallurgy and heat treatment in 1981.

Dr. Warmuzek is a research worker in the Metallography Laboratory at Foundry Research Institute in Kraków. Her research experience is in the areas of classical metallography, scanning electron microscopy and x-ray microanalysis, and quantitative metallography. The alloy microstructure formation (determining and modifying factors, modeling, and simulation) and intermetallic phases in aluminum alloys are the specialties of her scientific interests.

She has authored or coauthored more than 24 papers on the aforementioned topics. Dr. Warmuzek is the author of the article "Metallographic Techniques for Aluminum and Its Alloy" in the revised edition of *Metallography and Microstructures,* Volume 9, *ASM Handbook*, to be published in 2004.

Preface

This atlas of *microfractographs*—fracture images seen under a microscope, includes both profile and surface views of the specimens, and comprises a systematic documentation of aluminum-silicon alloys with relevant descriptions. The challenges of fractography are discussed in a comprehensive manner. The atlas contains images of fractures obtained during laboratory testing of mechanical properties. A set of images covers hypoeutectic, eutectic, and hypereutectic (mainly permanent mold) cast aluminum-silicon alloys. The surface of fractures visible under high-resolution scanning electron microscopy (SEM) are shown together with subsurface effects on metallographic specimens normal to the fracture plane, observed with light microscopy.

This book also deals with the physical, crystallographic, and microstructural aspects of the formation of aluminum-silicon cast alloys as they relate to mechanical properties. The criteria of fracture classification are established along with a set of the most important data on the morphology of the basic types of fractures that occur in metals used in structures. These may be used as a basis in the classification of fractures and in an assessment of the fracture path, its mechanism, and conclusions on the possible causes of a failure.

The alloys presented in this atlas are commercially important as their high strength-to-weight ratios make them suited for applications where reduction of weight is a design consideration, such as in automobile engine blocks, gear boxes, aerospace castings, and consumer products, as well as in marine and architectural uses.

This book is an English version of the atlas that was published in Poland in 2000. The author wishes to acknowledge her teacher Professor Stanisław Gorczyca, one of the pioneers of electron microscopy in Poland, to whom she is indebted for helping her finding a place in the world of technology. The author wishes to thank George Vander Voort, FASM, who brought this work to the attention of ASM International and Dr. Jerzy Tybulczuk (Foundry Research Institute) for support in the conception of this book.

Małgorzata Warmuzek

CHAPTER 1

Introduction to Aluminum-Silicon Casting Alloys

COMMERCIAL CAST ALUMINUM-SILICON alloys are polyphase materials of composed microstructure belonging to the Aluminum Association classification series 3*xx.x* for aluminum-silicon plus copper and/or magnesium alloys and 4*xx.x* aluminum-silicon alloys. They are designated by standards such as CEN EN 1706, "Aluminum and Aluminum Alloys. Castings. Chemical Composition and Mechanical Properties," and are designated in ASTM standards according to the method of casting:

- ASTM B 26/B 26M, "Specification for Aluminum-Alloy Sand Casting"
- ASTM B 85, "Specification for Aluminum-Alloy Die Casting"
- ASTM B 108, "Specification for Aluminum-Alloy Permanent Mold Casting"

Their use as structural materials is determined by their physical properties (primarily influenced by their chemical composition) and their mechanical properties (influenced by chemical composition and microstructure).

The characteristic property of aluminum alloys is relatively high tensile strength in relation to density (Table 1.1) compared with that of other cast alloys, such as ductile cast iron or cast steel. The high specific tensile strength of aluminum alloys is very strongly influenced by their composed polyphase microstructure.

The silicon content in standardized commercial cast aluminum-silicon alloys is in the range of 5 to 23 wt%. The structure of the alloys can be hypoeutectic, hypereutectic, or eutectic, as can be seen on the equilibrium phase diagram (Fig. 1.1a). The properties of a specific alloy can be attributed to the individual physical properties of its main phase components (α-aluminum solid solution and silicon crystals) and to the volume fraction and morphology of these components.

Table 1.1 Mechanical properties of selected cast engineering materials

Alloy	Ultimate tensile strength (UTS), MPa	Density (ρ), kg/m^3	Specific strength (UTS/ρ), m^2/s^2
Pure Al (99.9999% Al) Al (4N)	78	2699	0.03
Al-7%Si, T6	210	2685	0.09
Al-5%Si-2%Cu, T6	310	2690	0.12
Al-9%Si, T6	240	2650	0.10
Al-20%Si, T6	200	2650	0.08
Iron	1.9	7650	0.00024
Gray cast iron	380	7100	0.05
Ductile cast iron	900	7200	0.13
Austempered ductile cast iron	1200	7200	0.17
Cast carbon steel	650	7850	0.08
Cast stainless steel	880	7850	0.11

1.1 Properties of α-Aluminum Solid Solution

The α-aluminum solid solution is the matrix of cast aluminum-silicon. It crystallizes in the form of nonfaceted dendrites, on the basis of crystallographic lattice of aluminum. This is a face-centered cubic (fcc) lattice system, noted by the symbol *A*1, with coordination number of 12, and with four atoms in one elementary cell (Ref 1–3). Lattice *A*1 is one of the closest packed structures, with a very high filling factor of 0.74 (Fig. 1.2a). The plane of the closest filling is the plane {111}, and the direction <110> is the closest filling direction in this lattice (Table 1.2). Atoms are connected with metallic bonds characterized by isotropy and relatively low bonding energy (Ref 1, 2, 8). Each aluminum atom gives three valence electrons to an electron gas, filling the spaces among the nodes of the crystallographic lattice, formed by aluminum ions (Fig. 1.2b). Under external loading these ions can change their relative position in the lattice in some range, without breaking the interatomic bonds (Ref 1). The plastic deformation of crystals of metallic bonds is the macroscopic effect of this relative displacement of ions in nodes of their lattice. The breaking of the continuity of interatomic bonds in the ideal crystal takes place when the external stress exceeds the cohesive force value in the crystallographic planes (Ref 4–7). The value of this critical stress, estimated in Eq 1.1, is equal to *E*/10 (Ref 4):

$$\sigma_{max} = (2E\Gamma_s/\pi b)^{1/2} \quad \text{(Eq 1.1)}$$

where σ_{max} is the stress along axis perpendicular to crystallographic plane, in which the interatomic bonds are broken, *E* is the elastic modulus, Γ is the surface free energy, and *b* is the atomic diameter.

The value of this material constant is directly dependent on the physical properties of the crystallographic lattice and the atom size. On the basis of several different fracture models, it has been determined that the σ_{max} value can be in the range of $E/4$ to $E/15$ (Ref 5). The value of σ_{max} is called the theoretical tensile strength of ideal crystal. This value is several times greater than the tensile strength estimated experimentally for the real crystals or polycrystalline materials (Table 1.3). The tensile strength of the whisker crystals of aluminum can be compared to the theoretical one (their tensile strength is only 6.6 times less than theoretical one).

Theoretical proof stress—defined as the stress causing the permanent plastic deformation of crystallographic lattice (critical tangent stress τ_{max} on slip plane, $\tau_{max} = G/2$, where G is shear modulus)—is 10^4 times higher than the value estimated experimentally

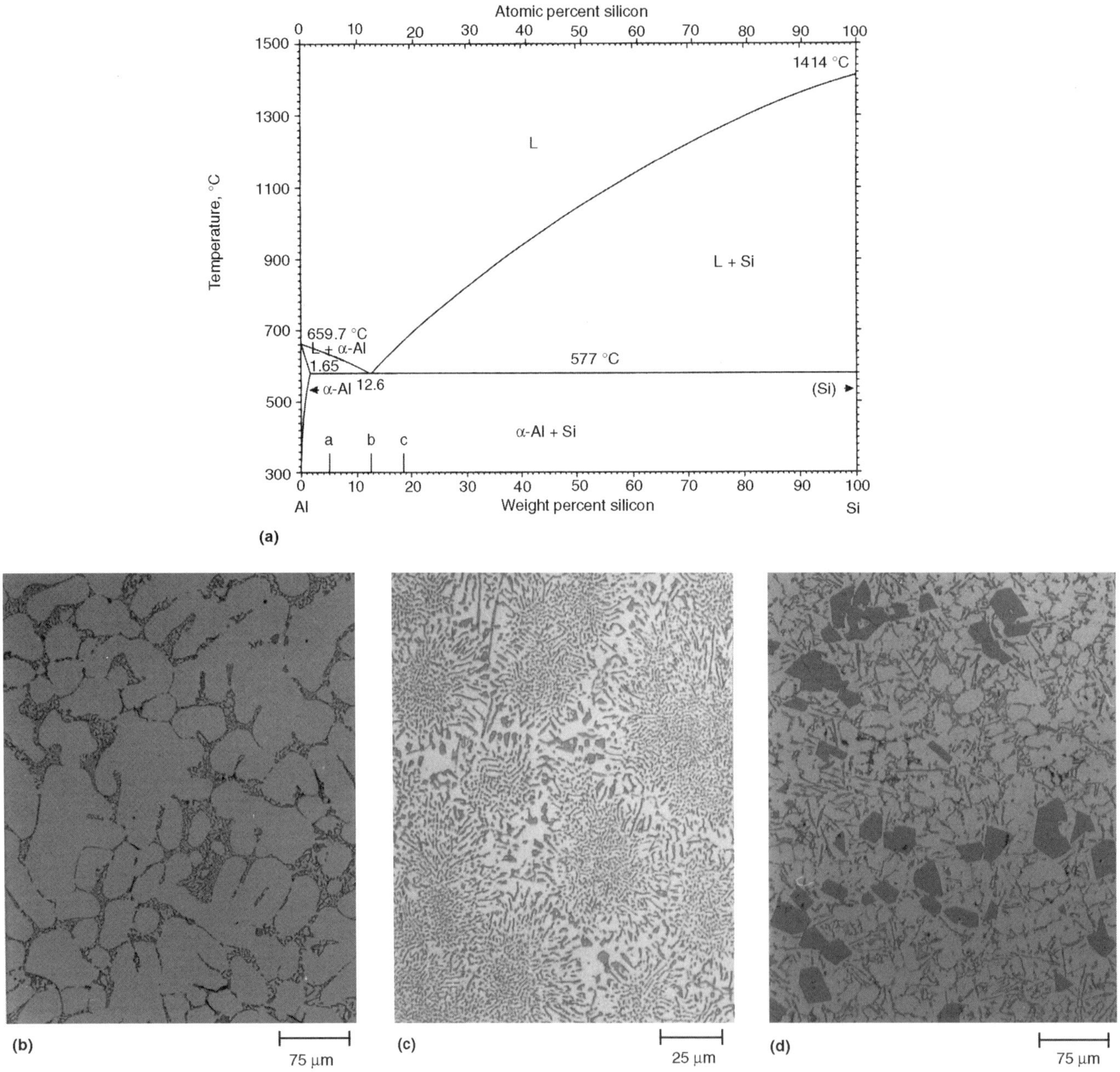

Fig. 1.1 Commercial cast aluminum-silicon alloys. (a) Al-Si equilibrium diagram. (b) Microstructure of hypoeutectic alloy (1.65-12.6 wt% Si). 150×. (c) Microstructure of eutectic alloy (12.6% Si). 400×. (d) Microstructure of hypereutectic alloy (>12.6% Si). 150×

Table 1.2 Parameters of the elementary cells of *A*1 and *A*4 crystal lattice

Element	Lattice type	Unit cell dimension, nm	Coordination number	Number of atoms in the unit cell	Filling factor	Bonding energy, kJ/mol (Ref 1)
Al	*A*1	0.40333	12	4	0.74	105–837
Si	*A*4	0.543035	4	8	0.34	523–1255

(Ref 5, 6). The reason for such comparatively low tensile strength in the real crystallographic lattice is due to the presence of defects, such as point defects (vacancies), line defects (dislocations), and surface defects (stacking faults).

The stacking-fault energy of crystallographic lattice of aluminum is very high, and very high density of moving dislocations (Ref 5, 6) is present. In *A*1 (fcc) metals, the Peierls-Nabarro (P-N) forces—that is, the resistance to dislocation movement—are low and almost do not affect the proof stress value. Their effect becomes quite noticeable at the liquid nitrogen temperature, −196 °C (−321°F) (Ref 5, 6).

The slip, causing permanent plastic deformation, is relatively easy, because in the *A*1 lattice there are 12 systems of easy slip: {111}, <110>. The plane {111} of the smallest surface energy is an energy-privileged plane of the easy slip. The small distance between partial dislocations makes their recombination easy, and the wave type transverse slip (of wavy glide type) can take place as well (Ref 6). The features of the real *A*1 lattice, mentioned above, cause the low resistance to deformation. It can be estimated from an empirical formula (Ref 6):

$$\sigma = k\varepsilon^n \quad \text{(Eq 1.2)}$$

where σ is real stress, ε is real strain, k is the strain-hardening factor, and n is a material constant.

The strain-hardening factor in Eq 1.2 for aluminum is 0.15 to 0.25, which is about half that of copper, bronze, or austenitic steel (Ref 6).

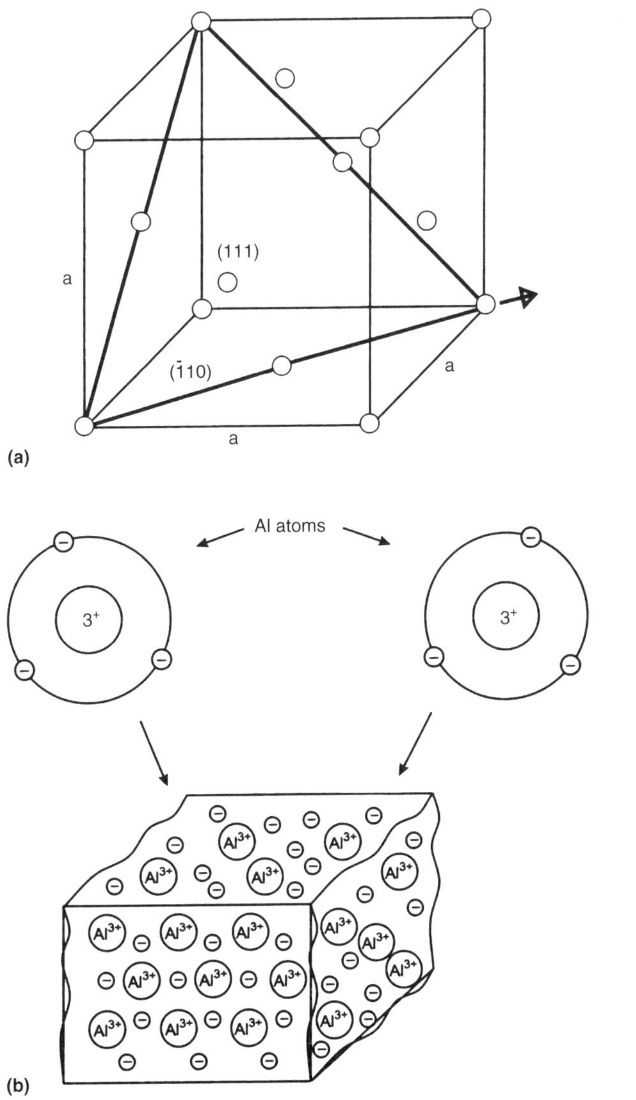

Fig. 1.2 Crystal structure of aluminum. (a) Elementary cell of a cubic crystalline lattice *A*1. (b) Aluminum atoms with outer electrons that develop the interatomic bonds in lattice *A*1. Source: Ref 1

1.2 Properties of Silicon Crystals

The silicon precipitates, present in commercial aluminum-silicon alloys, are almost pure, faceted crystals of this element (Fig. 1.3). They can have different morphology: primary, compact, massive precipitates in hypereutectic alloy or branched plates in aluminum-silicon eutectic (Ref 10–13).

Silicon crystal lattice is *A*4, cubic, of diamond type. Each atom is bonded with four others with covalent bonds, forming a tetrahedron. Eight tetrahedrons form one elementary cell of *A*4 lattice,

Table 1.3 Mechanical properties of aluminum and silicon crystals

Property	Al	Si
Peierls-Nabarro forces	Small	Big
Stacking fault energy	Big 250 mJ/m^2(a) 200 mJ/mol(b)	Small
Slip system	{111}; <110>(a)(c)	{111}; <1$\bar{1}$0>(a)
Strain-hardening factor	0.15–0.25(a)	...
Shear modulus (G) of monocrystal, GPa	26.7(d)(e) 29(c)	...
Shear modulus (G) of polycrystal, GPa	27.2(c)	40.5(c)
Theoretical yield strength (critical tangent stress on slip plane), GPa	4.26(e) 11.3(a)	6.4(c)
Experimental yield strength of monocrystal, MPa	0.78(d)(f)	...
Elastic modulus (E) of monocrystal, GPa	c_{11}, 108 c_{12}, 62 c_{44}, 28(g)	c_{11}, 166 c_{12}, 64 c_{44}, 79(g)
Elastic modulus (E) of whisker, GPa	506(c)	169(c)
Whisker tensile strength (R_m), GPa	15.5(c)	6.6(c)
Elastic modulus (E) of polycrystal, GPa	70(c) 71.9(c)	115(c)
Tensile strength of polycrystal, MPa	99.999% Al, 44.8(b) 99.99% Al, 45(h) 99.80% Al, 60(h)	5.3 GPa(c)
Hardness	99.999% Al, 120–140 HV(i)	1000–1200 HV(j) 8700–13500 N/m^2(a)
Cleavage energy	...	{111}, 890 mJ/m^2(c)
Point defects hardening factor	G/10 for symmetric defects; 2G for nonsymmetric defects(a)	...

Source: (a) Ref 6. (b) Ref 1. (c) Ref 7. (d) Ref 8. (e) Ref 2. (f) Ref 5. (g) Ref 9. (h) Ref 3. (i) Ref 10. (j) Ref 11 face centered, with four additional atoms from the center of each tetrahedron. This structure is less close packed than *A*1 lattice (Table 1.2). The filling factor is 0.34 (Ref 1, 8, 9). The neighboring atoms give four valence electrons and form a common hybrid orbital. The common, external shell circles the atoms in the lattice nodes and forms electron pairs of antiparallel spins (Fig. 1.4b) (Ref 2, 8). Characteristic features of the covalent bond are its high energy (523 to 1255 kJ/mol) and its anisotropy (Ref 1, 8, 9). Atoms, connected with covalent bonds, cannot displace under an external force until the bonds are completely broken. The material then cracks instantly, and decohesion takes place on the cleavage planes. Microscopic observations showed that the cleavage plane is the preferential plane for the brittle fracture, because of its small surface energy. In silicon this is plane {111}. Cleavage work, necessary for breaking the atomic bonds in this plane, is equal to 890 J/m^2 (Ref 7). In crystals with covalent bonds, the density of dislocations is small, and P-N forces are high (Ref 1, 5, 6). This is the reason for the high proof stress of silicon and its inclination to brittle cracking. Its slip system, which can be active in silicon crystals, is {111} $<\bar{1}10>$ (of planar glide type) (Ref 6).

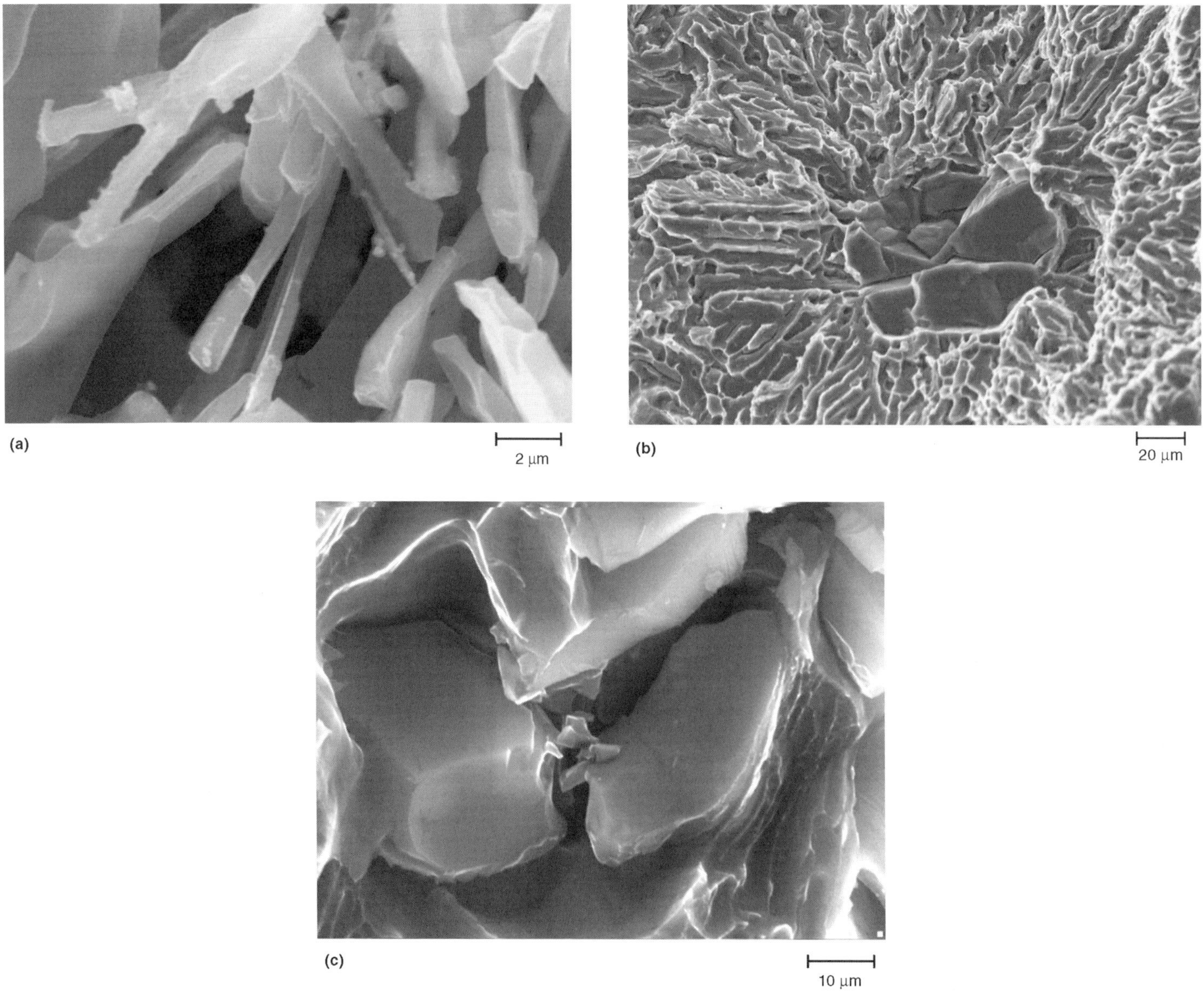

Fig. 1.3 Morphology of the silicon crystals in aluminum-silicon alloys. (a) Silicon crystals in eutectic as-cast alloy. Scanning electron micrograph (SEM). 6500×. (b) Primary silicon crystals in hypereutectic as-cast alloy. SEM. 400×. (c) Silicon crystals in hypoeutectic alloy modified, after heat treatment. SEM. 1500×

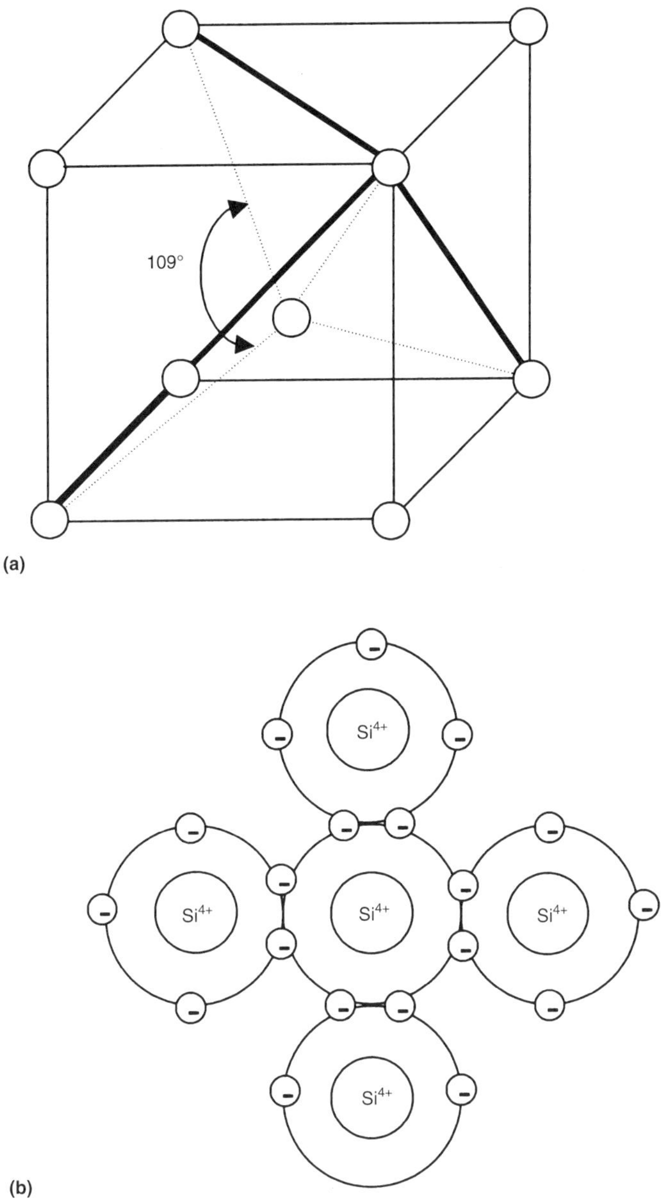

Fig. 1.4 Crystal structure of silicon. (a) Elementary cell of the crystalline lattice silicon. (b) Interatomic bonds in lattice silicon. Source: Ref 1, 8, 9

1.3 Properties of Aluminum-Silicon Alloys

The simplest model of microstructure of cast aluminum-silicon alloys can be presented in the form shown in Fig. 1.5: a soft continuous matrix (α-aluminum-solid solution) containing hard precipitates of silicon of different morphology.

Assuming an important simplification, the average stress in this material can be evaluated as a linear function of the volume fraction of silicon (Ref 7):

$$\sigma = \sigma_\alpha V_V^\alpha + \sigma_{Si} V_V^{Si} = \sigma_\alpha + V_V^{Si} (\sigma_{Si} - \sigma_\alpha) \quad \text{(Eq 1.3)}$$

where σ_α and σ_{Si} are stresses in the volume unit.

The stress-intensity factor depends on the elastic and the plastic properties of the matrix and on the size of the brittle phase particles (Ref 14, 15). In Eq 1.3, the influence of the morphology, the average size, and the distribution of brittle particles, that is, silicon precipitates, are not taken into account. These microstructure parameters can differentiate the properties of materials of similar value of the silicon volume fraction to an important degree. In the polycrystalline material, the Hall-Petch equation can express the influence of the morphology of the microstructure constituents on the proof stress (Ref 1, 6, 7):

$$\sigma_{pl} = \sigma_s + k_m d^{-1/2} \quad \text{(Eq 1.4)}$$

where σ_{pl} is the proof stress of the polycrystalline specimen, σ_s is the resistance of the lattice to dislocation movement, k_m is the hardening factor (effect of hardening by grain boundaries), and d is the grain diameter.

The stress σ_s can be divided into two parts: σ_d and σ_p. σ_d is independent of temperature but dependent on the structure of lattice, and it expresses interaction among dislocations, precipitates, and additional atoms. σ_p is temperature dependent and is connected with P-N stress value (Ref 6):

$$\sigma_{pl} = \sigma_p + \sigma_d + k_m d^{-1/2} \quad \text{(Eq 1.5)}$$

where σ_p represents P-N stresses, a short-range effect (<1 nm); σ_d is the dislocation stress field, a medium-range effect (10 to 100 nm); and $k_m d^{-1/2}$ is the long-range effect (>1000 nm). One can say that the influence of the degree of microstructure dispersion on the proof stress is a long-range effect.

Many published experimental examinations show that in case of dendrite structure materials the microstructure effect in the Hall-Petch formula is dependent on λ, the dendrite arm size, and γ, the size of silicon lamellas (Ref 16–20). Relationships between ultimate tensile strength (R_m) and secondary dendrite arm size, evaluated experimentally for alloy C356, can be expressed by:

$$R_m = k + k_2 \gamma^{-1/2} + k_3 \lambda^{-1/2} \quad \text{(Eq 1.6)}$$

$$R_{0.2} = k + k_5 \gamma^{-1/2} + k_6 \lambda^{-1/2} \quad \text{(Eq 1.7)}$$

where k, k_2, k_3, k_5, and k_6 are empirical constants, γ is the size of silicon lamellas in interdendritic eutectic regions, and λ is the secondary dendrite arm size. $R_{0.2}$ is the 0.2% proof strength.

The mechanical properties of cast aluminum-silicon alloys can be improved by cast technology and heat treatment processes that:

- Increase the strength of soft matrix
- Decrease the brittle fracture risk in the polyphase regions
- Increase the degree of dispersion of the dendritic structure

An increase in strength of soft matrix of α-aluminum solid solution can be achieved by its hardening with point defects, such as sub-

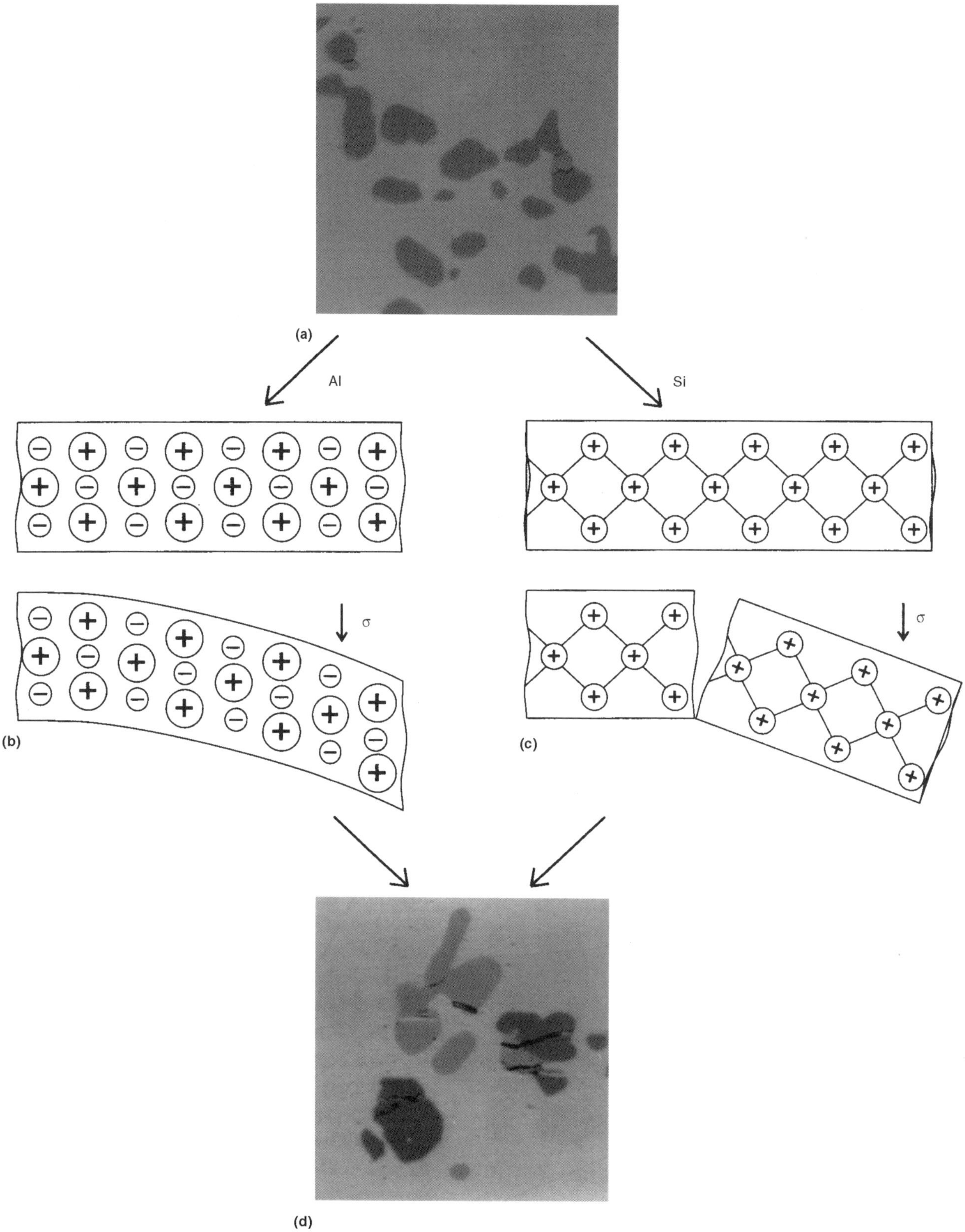

Fig. 1.5 Properties of an aluminum-silicon alloy formed by its phase components. (a) Biphase microstructure of the aluminum-silicon brittle hard silicon particles in soft plastic aluminum matrix. (b) Reaction of the matrix under external loading (according to Ref 1). (c) Reaction of silicon particle under external loading (according to Ref 1). (d) Result of the loading of material

stituted atoms and vacancies or by precipitation hardening with dispersion particles of the second phase.

Decreasing the risk of brittle fracture in polyphase regions can be realized only by reducing the intensity of the stress-concentration effect on the silicon particles and by eliminating microregions of potential crack initiation. The breaking of the silicon precipitates network and their spheroidization are very important. This allows a decrease in the stress-concentration factor value in these regions, depending on brittle phase morphology (Ref 6, 14):

$$K_n = 2a/b \qquad \text{(Eq 1.8)}$$

where a is one-half the length of a particle and b is one-half the width of a particle.

Material of fine microstructure morphology has less tendency for low-energy brittle cracking. To obtain acceptable morphology and degree of dispersion of the components, the foundry personnel can interfere at the crystallization stage by modifying the alloy and by controlling the solidification path and then, in the solid state, by heat treatment.

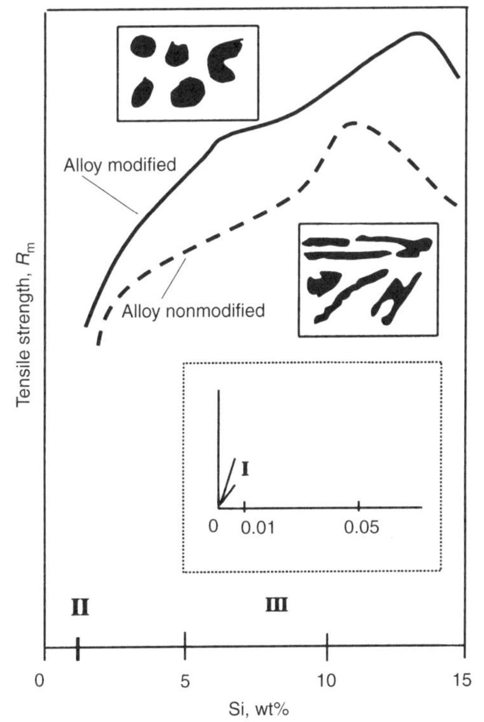

Fig. 1.6 Tensile strength versus silicon content in aluminum-silicon cast alloy. Source: Ref 1

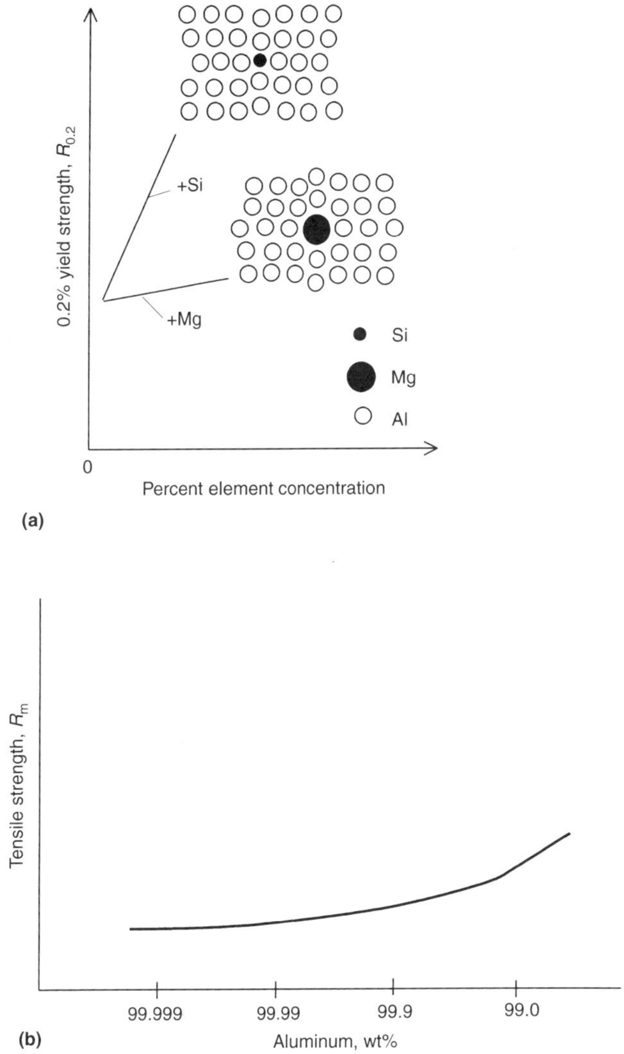

Fig. 1.7 Influence of addition elements on mechanical properties of aluminum. (a) Deformation of crystal lattice caused by substitution atoms. (b) Change of the mechanical properties of α-aluminum solid solution in presence of addition atoms. Source: Ref 1, 10

Table 1.4 Properties of the alloying elements in aluminum-silicon commercial alloys

Element	Atomic number, Z	M	Crystal symmetry	Unit cell parameters α, nm	Unit cell parameters c, nm	Atomic radius nm	Ion radius nm	Density g/cm^3	Melting point, °C	Max. solubility in α-Al, wt%
Mg	12	24.32	*A*3 hexagonal	0.320	0.520	0.160	0.066	1.74	650	14.9(450 °C)
Al	13	26.97	*A*1 cubic	0.4041	...	0.143	0.051	2.699	660.5	...
Si	14	28.06	*A*1 cubic	0.543	...	0.118	0.042	2.35	1440	1.65 (577 °C)
Mn	25	54.93	*A*1 cubic	0.893	...	0.112	$0.080^{(+2)}$ $0.066^{(+3)}$	7.43	1240	1.82 (660 °C)
Fe	26	55.84	Cubic	*A*1, 0.369 (916 °C) *A*2, 0.288 (20 °C)	...	*A*1, 0.124 *A*2, 0.127	*A*1, 0.064 *A*2, 0.074	7.87	1538	0.052 (655 °C)
Cu	29	63.57	*A*1 cubic	0.361	...	0.128	...	8.96	1083	5.67(550 °C)

Source: Ref 1, 3, 8–10

1.4 Effects of Different Levels of Silicon Contents

In the equilibrium diagram presented in Fig. 1.1 and in Fig. 1.6, several characteristic ranges of silicon content can be identified. In each range (I, II, III of Fig. 1.6), a different mechanism of the silicon influence on the properties of the alloy is present.

1.4.1 *Silicon Contents of 0 to 0.01 wt%*

In range I, silicon is substituted for aluminum atoms in solid solution. The silicon atoms situated in the nodes of crystallographic lattice strengthen the α-aluminum solid solution (Table 1.4). Deformation of lattice caused by the difference in diameter of aluminum and silicon atoms makes the dislocations movement difficult. The atoms of other alloying elements can act in a similar manner. Even though their solubility in α-aluminum solid solution is very small, trace quantities of alloying elements can change the mechanical properties of aluminum-silicon alloy to an important degree (Fig. 1.7) (Ref 1, 3, 10, 11). This is caused by very strong interaction between either screw and edge dislocations and by the stress field introduced by the substitution atoms. Range of an introduced misfit in the hydrostatic stress field is directly proportional to the difference of atom radii between the matrix and the additional elements. The effect of the strengthening of α-aluminum solid solution by the substitution of atoms of smaller radii than aluminum (such as silicon, manganese, iron, and copper) is more evident than in the case of atoms of larger radii (such as magnesium) (Fig. 1.7a). Some level of the strengthening can also be achieved owing to the nonsymmetric stress field around the disk vacancies, interacting with the nonsymmetric part of the stress field of either screw or edge dislocations (Table 1.3).

1.4.2 *Silicon Contents of 0.01 to 1.65 wt%*

In range II, a temperature-dependent terminal solid solution of silicon in aluminum forms can be strengthened by dispersed precipitation. During fast cooling α-aluminum solid solution can be supersaturated and then, as a result of its tendency to achieve the thermodynamic equilibrium, the dispersed particles of silicon pre-

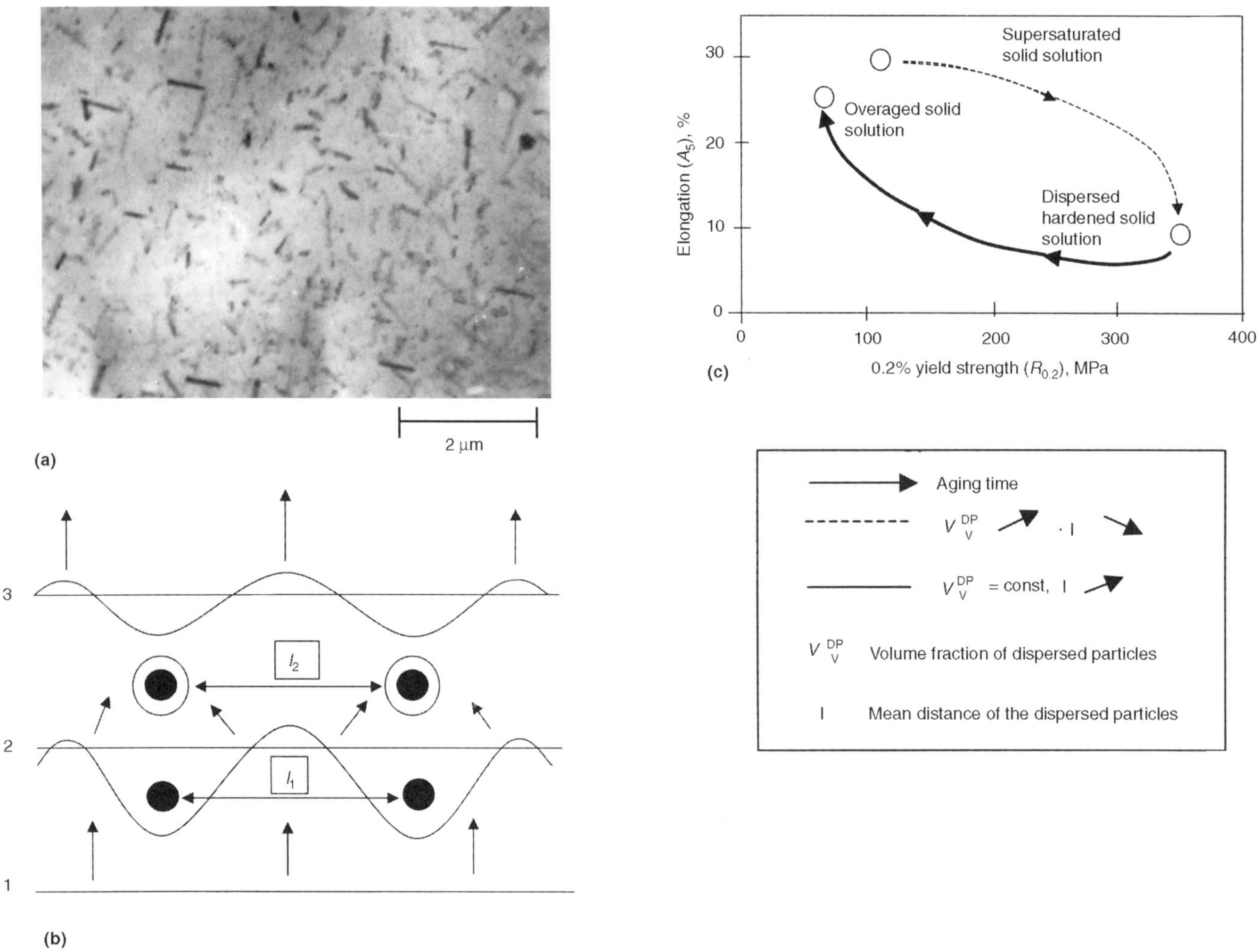

Fig. 1.8 Precipitate hardening of the supersaturated α-Al-solid solution. (a) Morphology of the disperse precipitates, C355-T6 alloy. TEM, 10,000×. (b) Orowan mechanism—dislocation displacement in matrix with hard disperse particles. Areas labeled 1, 2, and 3 of (b) show successive steps of the process of displacement of the dislocation through material among dispersed precipitates. l_1, initial distance between precipitates. l_2, apparent distance when the first dislocation has gone through. (c) Change of material properties depending on the particles morphology. Source: Ref 1, 6, 12

cipitate on {111} and {100} planes (Ref 10). A similar event takes place in the presence of copper, manganese, and magnesium atoms. Dispersed particles of intermetallic phases can also precipitate from a supersaturated α-aluminum solid solution. The material hardening with such particles can be explained taking into account an Orowan model (Fig. 1.8). Shear stress, which causes particle lateral dislocation, can be expressed by (Ref 5):

$$\tau = G\mathbf{b}/l \qquad \text{(Eq 1.9)}$$

where τ is shear stress, G is shear modulus, $\mathbf{b}$ is Burgers vector, and l is the distance between dispersed particles.

As the distance between particles decreases and achieves some critical value, with simultaneous enlargement of their size, the stress necessary to move dislocations increases and material becomes hardened.

1.4.3 Silicon Contents Greater Than 1.65 wt%

In this silicon concentration range (III), the two-phase alloys solidify and the influence of silicon on properties can be described by Eq 1.3. Some misfits from linear dependence, visible in Fig. 1.6 (Ref 1), reflect the influence of morphology and distribution of silicon precipitates.

REFERENCES

1. D.R. Askeland, *The Science and Engineering of Materials,* PWS-Kent Publishing Co., 1987
2. J. Massalski, *Fizyka dla inżynierów* (*Physics for Engineers*), Vol 2, Wyd. Naukowo-Techniczne, Warsaw, 1976 (in Polish)
3. J.E. Hatch, Ed., *Aluminum: Properties and Physical Metallurgy,* American Society for Metals, 1984
4. M.F. Ashby, C. Ghandi, and M.R. Taplin, Fracture-Mechanism Maps and Their Construction for FCC Metals and Alloys, *Acta Metall.,* Vol 27, 1979, p 699–729
5. G.E. Dieter, *Mechanical Metallurgy,* 3rd ed., McGraw-Hill, 1986 p 241–271
6. R.W. Hertzberg, *Deformation and Fracture Mechanics of Engineering Materials,* John Wiley & Sons, 1989
7. M.L. Bernsztejn and W.A. Zajmowskij, *Struktura i własności mechaniczne metali (Structure and Mechanical Properties of Metals),* Wyd. Naukowo-Techniczne, Warsaw, 1973 (in Polish)
8. L. Kalinowski, *Fizyka metali (Physics of Metals),* PWN, Warsaw, 1973 (in Polish)
9. C. Kittel, *Solid State Physics,* John Wiley & Sons, 1957
10. L.F. Mondolfo, *Aluminium Alloys: Structure and Properties,* Butterworths, London-Boston, 1976
11. Z. Poniewierski, *Krystalizacja, struktura i własności siluminów (Crystallization, Structure and Properties of Silumins),* Wyd. Naukowo-Techniczne, Warsaw, 1989 (in Polish)
12. J.R. Davis, Ed., *ASM Specialty Handbook: Aluminum and Aluminum Alloys,* ASM International, 1993
13. S.-Z. Lu and A. Hellawell, Modification of Al-Si Alloys: Microstructure, Thermal Analysis and Mechanics, *JOM,* Vol 47 (No. 2), 1995, p 38–40
14. A. Gangulee and J. Gurland, On the Fracture of Silicon Particles in Al-Si Alloys, *Trans. Metall. Soc. AIME,* Vol 239 (No. 2), 1967, p 269–272
15. M.A. Przystupa and T.H. Courtney, Fracture in Equiaxed Two Phase Alloys: Part I. Fracture with Isolated Elastic Particles, *Metall. Trans. A,* Vol 13A (No. 5), 1982, p 873–879
16. H. Arbenz, Qualitatsbeschreibung von Aluminium—Gusstucken Anhand von Gefugemerkmalen (The Use of Structural Features to Determine the Quality of Aluminum Castings), *Giesserei,* Vol 66 (No. 19), 1979, p 702–711 (in German)
17. C.H. Caceres and Q.G. Wang, Dendrite Cell Size and Ductility of Al-Si-Mg Casting Alloys, *Int. J. Cast Metals Rev.,* Vol 9, 1996, p 157–162
18. O. Vorren, J.E. Evensen, and T.B. Pedersen, Microstructure and Mechanical Properties of AlSi(Mg) Casting Alloys, *AFS Trans.,* Vol 92, 1984, (84-462), p 549–466
19. H.M. Tensi and J. Hogerl, Metallographische Gefuge—Untersuchungen zur Qualitatssicherung von Al-Si Gussbauteilen (Metallographic Investigation of Microstructure for Quality Assurance of Aluminum-Silicon Castings), *Metall,* Vol 48 (No. 10), 1994, p 776–781 (in German)
20. P.N. Crepeau, S.D. Antolovich, and J.A. Warden, Structure-Property Relationships in Aluminum Alloy 339-T5: Tensile Behavior at Room and Elevated Temperature, *AFS Trans.,* Vol 98, 1990, p 813–822

CHAPTER 2

Fractography

FRACTOGRAPHY is a part of materials science that involves describing the topography of a separation surface formed during a breakage of the material continuity. Descriptions of the characteristic features of fracture surface and their classification are very important for establishing the dependence between the decohesion mechanism (dependent on physical and mechanical properties) and material microstructure (determined by chemical composition and production technology).

Fractographic examination of metals is used in metal science to (Ref 1–4):

- Evaluate the cause of material destruction by revealing and identifying internal discontinuities such as internal cracks, porosity, inclusions, and chemical or microstructural inhomogeneities
- Determine the decohesion mechanism by describing and classifying the characteristic morphological features of the fracture surface
- Estimate the stress field acting during decohesion by analyzing fracture morphology, taking into account both fracture surfaces
- Evaluate the degree of deformation on the crack path by the selected areas electron channeling (SAEC) pattern method
- Identify crack paths

2.1 Methods of Fracture Investigation

2.1.1 Fracture Surface Observation Using the Light Microscope

The light microscope has a limited application for observation and identification of the fracture surface because of its small depth of field and low resolution, compared with the electron microscope (Table 2.1). Nevertheless, in some cases the stereo light microscope can be used to identify structure defects such as macroporosity and slag inclusions revealed on the fracture surface or for examination of the fatigue lines and the range of crack zones. The light microscope is a useful tool to make fracture profile observations, on specially prepared metallographic microsections (Ref 1–4).

2.1.2 Fracture Surface Observations by Electron Microscopy

Transmission Electron Microscopy (TEM). Very careful preparation of the TEM specimen is necessary for investigation of fracture surfaces to obtain satisfactory contrast, because of the demands concerning the specimen thickness. The fracture surface observations are carried out mainly on thin, two-stage replicas (Fig. 2.1). Often, these are carbon replicas in the form of amorphous foils, obtained by covering a plastic negative of the fracture surface with a carbon film, evaporated in vacuum, in an electric arc. Improvement in the contrast of the replica can be achieved by shadowing with metal such as gold, platinum, chromium, silver, and palladium (metal vapors are settled obliquely from the source toward the replica surface). It should be noted that the resolution of the carbon replica is lower than the resolution of a high-resolution electron microscope. The direct one-step replicas are not used because replicas are fragile and difficult to remove from highly developed (rough) fracture surfaces. The separation of a replica from a fracture demands either chemical or electrolytic etching of a specimen, which can change the fracture topography irreversibly (Ref 1–7).

The scanning electron microscope (SEM) allows an indirect observation of the fracture surface in all ranges of magnification (Ref 1–3, 6–8). The large depth of field of an SEM is a very important benefit for fractographic investigations. Fracture surfaces can be observed with an SEM almost without any special preparation; nevertheless, the specimen should be clean. Cleaning can be done mechanically by rinsing in ultrasonic cleaner or in chemical reagents or electrolytes. The two last methods are used where the specimen surface is oxidized to such a degree that the oxide film changes surface topography or gathers electrostatic

Table 2.1 Examination possibilities of the light microscope, the scanning electron microscope, and the transmission electron microscope

Microscope	Information carriers	Resolution, nm	Depth of focus	Observation field
Light	Electromagnetic waves of length in range of visible light	>300	Small	Fracture surface directly; fracture profile via metallographic microsection
Electron transmission	Electron beam, electromagnetic waves	2–3	Small	Fracture surface indirectly (replicas)
Electron scanning	Electron beam, secondary electrons	6–9	Large	Fracture surface directly

charge, making the observation impossible, since a necessary condition for investigation by means of SEM is electric conductivity of the specimen, at least in the surface layer.

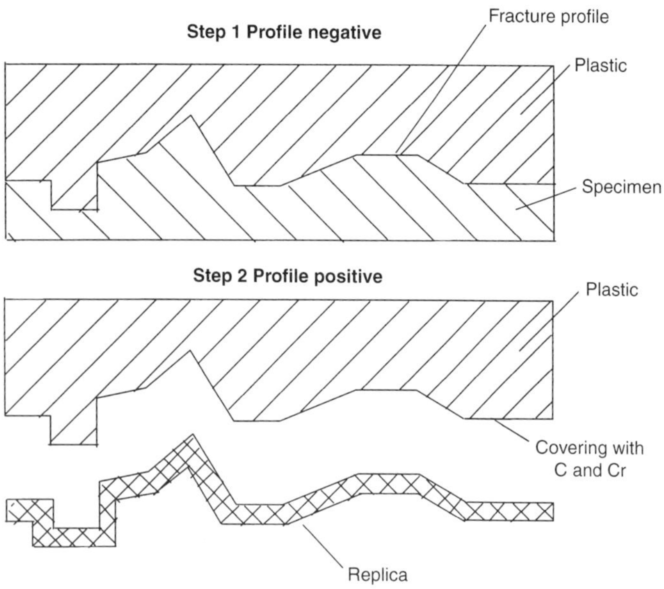

Fig. 2.1 Two-step plastic replica, successive stages of the preparation

Comparison of the methods of the fracture surface observation mentioned previously is presented in Table 2.1.

2.2 Qualitative Fractography

Fracture topography can be described on the basis of observations of its profile or surface. The identification and the systematization of the characteristic features of the fracture profile or the surface morphology, in conjunction with material properties and type of loading, can be the basis of the fracture classification.

2.2.1 Criteria for Fracture Classification

The main factors that form fracture morphology in engineering materials are the relations of external loading, cohesive force levels, and internal stresses in the crystallographic lattice at the different ranges of interaction (see Eq 1.4 and 1.5 in Chapter 1).

All the characteristic elements of the fracture morphology are a result of processes occurring in loaded material, such as the breaking of atomic bonds and the displacement of the atom positions in crystallographic lattice. Microscopic observations, using the electron microscope, allow one to distinguish a set of morphological features of the fracture surface, characteristic for different fracture mechanisms and materials (Fig. 2.2, Ref 9). Fracture morphology, under cyclic loading, differs from those observed for static or dynamic loading, so fatigue fractures form a separate group (Ref 1).

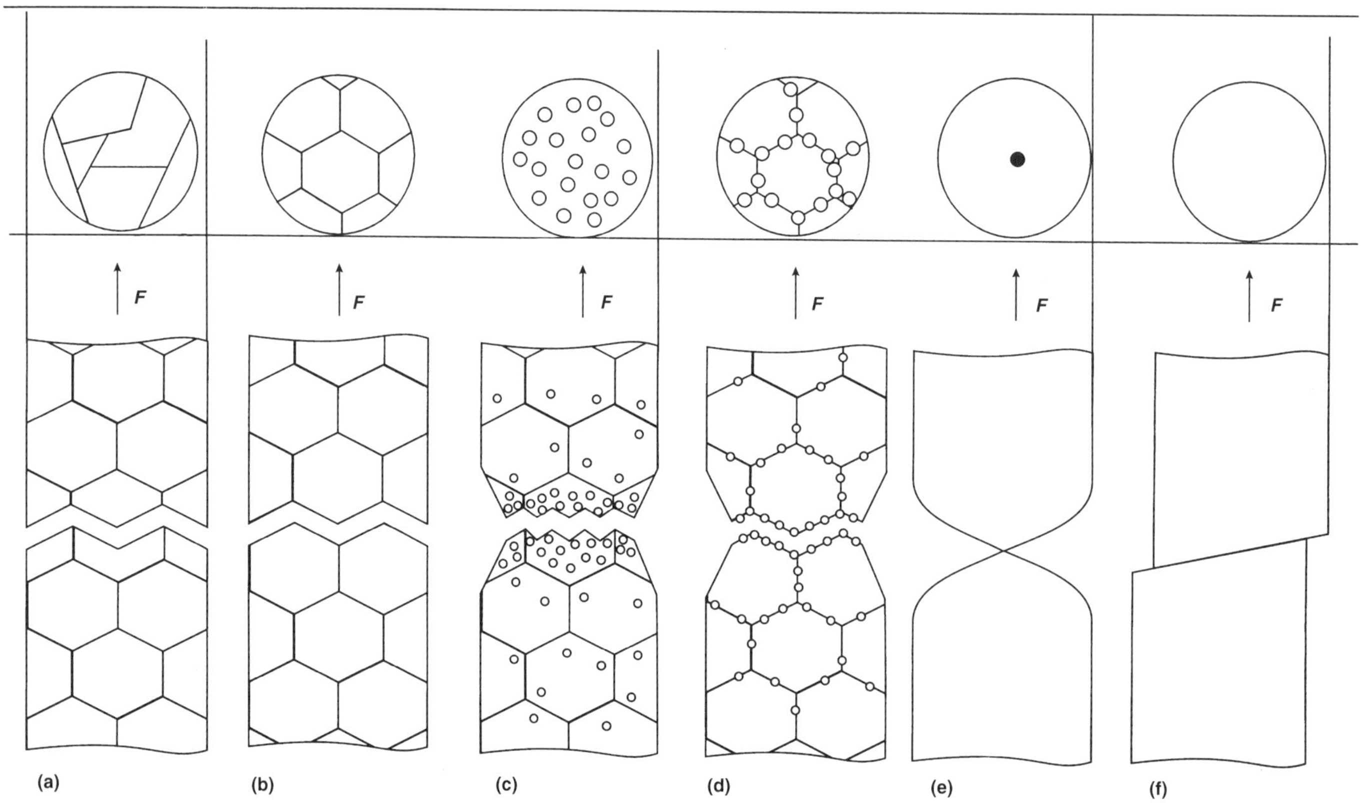

Fig. 2.2 Classification of fractures formed in polycrystalline materials during tensile testing ***F***, force. (a) Brittle transcrystalline. (b) Brittle intercrystalline. (c) Ductile transcrystalline. (d) Ductile intercrystalline. (e) Plastic. (f) By shear. Source: Ref 9

A single, general criterion for fracture classification is difficult to formulate. The fracture surface classification can be done on the basis of the following four main criteria (Ref 2, 4, 9–12):

Criterion	Variations
Fracture path	Transcrystalline fracture Intercrystalline fracture
Fracture energy	High energy Low energy
Mechanism of decohesion	Cleavage fracture Slip fracture
Material deformation	Plastic fracture Shear fracture Ductile fracture Cleavage fracture Mixed fracture

2.2.2 Low-Energy Fractures without Significant Deformation of the Crystal Lattice

Low-energy decohesion, without visible plastic deformation on the macroscopic scale, is the mechanism of formation of brittle fracture. In the crystallographic lattice of the brittle material, besides crystal defects, some microcracks are very often present. In these microregions, the stress concentration takes place and an initiated crack can propagate before the main external load exceeds the value of the material cohesion forces (Ref 9). The crack propagation is, in this case, very fast. It is estimated as equal to 0.7 times the speed of sound in the material. When the fracture travels along grain boundaries, it is called intercrystalline fracture; when it crosses the grains, it is called transcrystalline fracture (Ref 1, 2, 9, 12).

Transcrystalline brittle fracture is also called cleavage fracture. In this case, the cleavage mechanism of decohesion is active. The new, separated surfaces in material form by breaking of the atomic bonds in crystallographic lattice, without previous change in their relative position. The crack propagates along the crystallographic planes, named the cleavage planes. They are characterized by the close packing with atoms, and they have relatively low surface energy (Ref 1, 2, 12, 13). In Al (fcc) metals, this kind of fracture is not often observed. The cleavage plane is also not clearly defined.

A low-energy cleavage fracture is especially privileged at low temperature (below 0 °C). The materials ductile at the ambient temperature can crack in brittle mode at subzero temperatures; this means that mechanism of decohesion by slip is transformed into a cleavage mechanism. The surface of the cleavage fracture is in macroscopic scale weakly developed. In extreme situations, the ideal cleavage fracture surface in polycrystalline material should be a plane surface, smooth in atomic scale in each grain (crystallite) (Fig. 2.3–2.7).

Intercrystalline fracture occurs when the interface cohesion forces on the grain boundaries are weak. If cohesion forces in these zones become lower than cohesion forces on the cleavage planes or if there is not a sufficient number of slip systems to propagate the continuous plastic deformation in the successive grains of the loaded material, decohesion takes place along grain boundaries. In this manner, formation of the new separation surface will demand less energy.

Intercrystalline fracture is very often formed when chemical or structural segregation is present on the grain boundaries, as with precipitates or impurities, and can also be stimulated by thermal or corrosion factors.

Intercrystalline fracture belongs to the group of brittle fractures (Fig. 2.2b), but sometimes traces of plastic deformation can be observed on its surface (Fig. 2.2d), primarily at the nucleation and coalescence of voids in the neighborhood of the brittle precipitates (Ref 1, 2, 12, 13).

2.2.3 Fracture with Deformation of the Crystal Lattice

The formation of this kind of fracture demands higher energy input compared with brittle cracks. The energy is absorbed during plastic deformation, causing an activation of the slip systems in successive microregions (Ref 1, 2, 12, 13). As a result of this process, the characteristic dimples nucleate, grow, and join on the formed separation surface (Fig. 2.2c). This process starts with nucleation of small discontinuities (voids). The initiation of this process usually takes place on the interfaces between hard dispersed precipitates and the matrix. It can be also observed in the microdiscontinuities as micropores and sometimes on the surfaces of microcracks in hard precipitates or inclusions.

Generally, each region of the lower, interatomic or interface, cohesion in material is the preferred place for ductile fracture initiation. Under the triaxial stress state, before the crack front, the microvoids enlarge and join, forming dimples. This process is visible in macroscale as a plastic flow and can be intensified by an additional successive loss of cohesion on the interfaces. The presence of dimples on its surface classifies the fracture as a ductile fracture. The mechanism of the material deformation is by slip. Sometimes, in high-plasticity material, the nucleation and coalescence of the dimples is not possible, and, in this situation, it will flow until cracking is complete. The material in the macroscale elongates in the direction of the external stress and then flows until cracking at the point of reduced area (pinpoint mechanism). The factors limiting the process of nucleation and growth of microvoids are: high purity of material, high level of cohesion on interfaces, or fast relaxation of stresses in regions of their local concentration (Ref 9, 10). The characteristic feature of macroscopic deformation of high plasticity material is point necking (Fig. 2.2e, Ref 9). In the microscopic scale, one can observe, in the plastic microstructure constituents, the local effect of plastic flow in the form of tear ridges or micronecks.

Fracture by shear is formed as a result of the plane stress state (Ref 1, 2). The surface of the ideal shear fracture should be the plane field of the relative displacement of two slip planes. In real material, the surface of a shear fracture is usually more developed. It is caused by some heterogeneity of the strain stress, the faults of the crystalline lattice, and the presence of disperse particles. The microscopic observations allowed the classification of the following kinds of shear fracture:

- Plain fracture (Ref 11)
- Waved fracture (Ref 11)
- Fracture with shear dimples (Ref 2, 9–12, 14)

The last one is included in group of the ductile fractures. Complex process of its formation consists of the following stages: microvoid initiation (Fig. 2.8a), plastic flow, formation of dimples by void coalescence (Fig. 2.8b), and shear of dimples (Fig. 2.8d, f).

Mixed fracture surface is characterized by the simultaneous presence of the features of the brittle, plastic, and ductile fracture (Ref 1, 2, 5, 15).

Plastic-brittle fracture takes place when the features of brittle cracks and plastic deformation can be visible inside one grain of the same phase. The cleavage planes are not clearly limited by the field of one grain, but they are smaller, and often the tear ridges, present in the vicinity of the dimples, divide them (Fig. 2.9).

Cellular fracture is typical of polyphase material, where the microstructure components have different mechanical properties. On the cellular fracture surface, the features of both brittle and ductile fracture are present simultaneously. Each particular phase constituent cracks according to its proper decohesion mechanism. The areas of specific fracture morphology formed in this manner

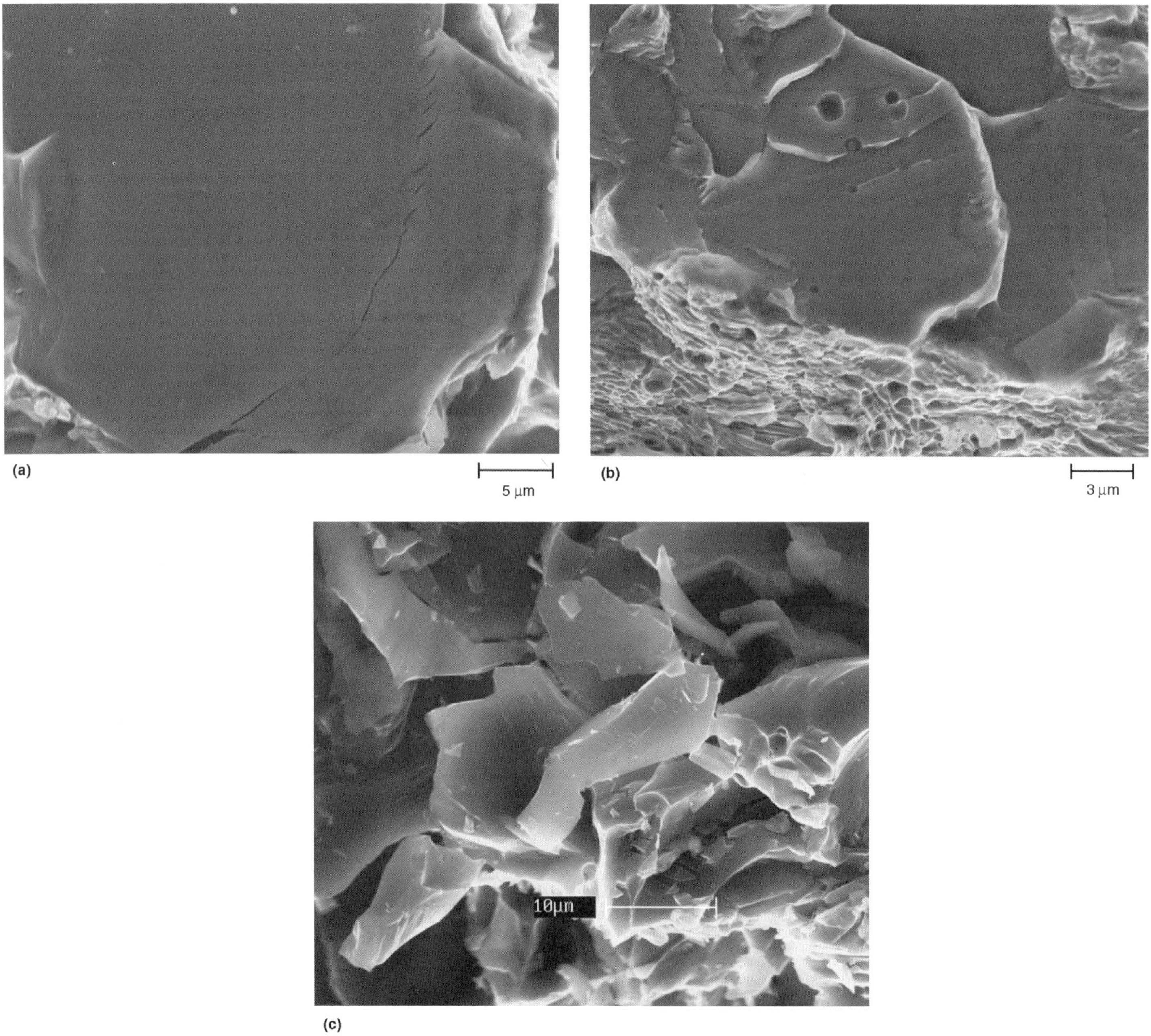

Fig. 2.3 Transcrystalline fracture, cleavage facets. (a) Smooth cleavage facets and secondary cracks, primary silicon in hypereutectic aluminum-silicon alloy, static tensile test. Scanning electron microscopy (SEM). 2800×. (b) Cleavage fracture on parallel cleavage planes, primary silicon in hypereutectic aluminum-silicon alloy, static tensile test. SEM. 3500×. (c) Cleavage fracture, primary silicon cracked on the several cleavage planes in hypereutectic aluminum-silicon alloy, static tensile test. SEM. 2000×

are separated by the phase boundaries. Very often in the boundaries between brittle and ductile phase, the continuity is preserved (Fig. 2.10).

2.2.4 *Description of the Fracture Profile*

Very important information concerning fracture path and microstructure components involved in crack process can provide an observation of the fracture profile, visible on specially prepared metallographic microsection, cut out perpendicularly to the fracture macrosurface (Fig. 2.11, 2.12). An observation, carried out by a light microscope, allows identification of the structure components crossed by the crack front, the structure components present in zone of material beneath fracture surface, and also estimation of the level of the interface cohesion forces on the grain boundaries or the interfaces. The length and position of the secondary cracks can also be revealed.

In Fig. 2.11, a sketch of a typical fracture profile for the polyphase material, of the microstructure model characteristic for Al-Si alloy, is presented.

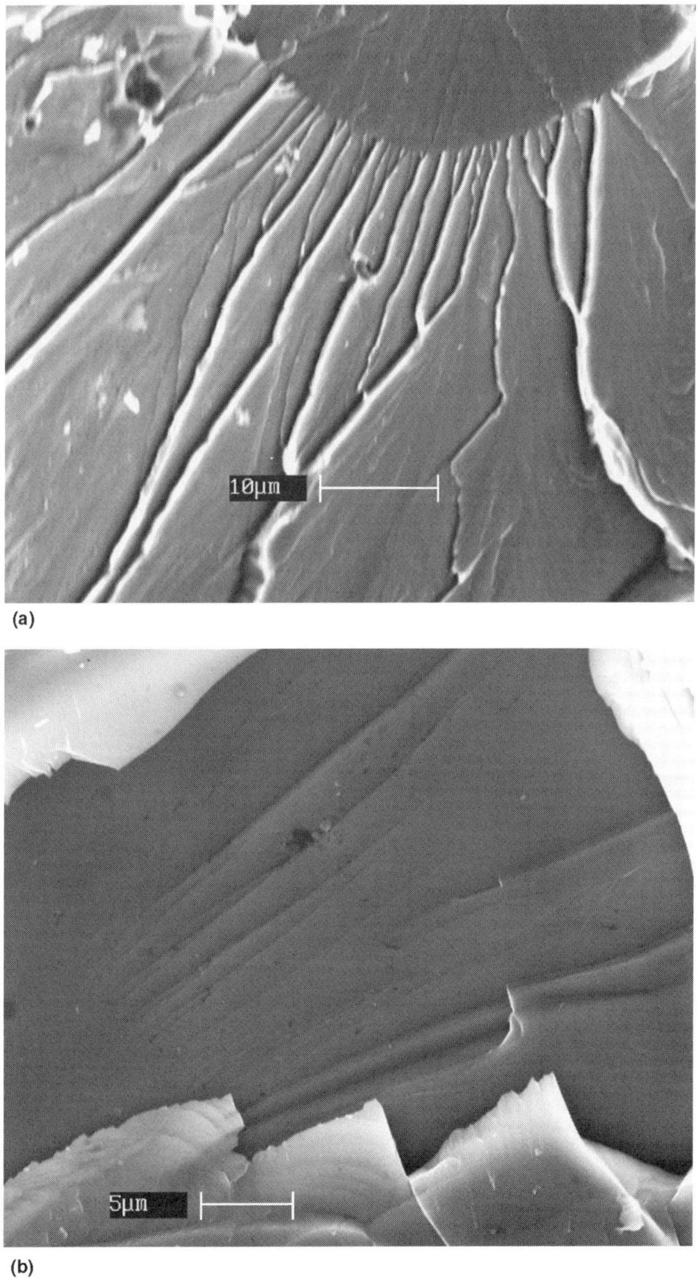

Fig. 2.4 System of steps forming rivers and river patterns on the cleavage fracture surface. (a) Ferrite grain in cast steel, impact test at −160 °C. SEM. 1800×. (b) Primary silicon in hypereutectic aluminum-silicon alloy, static tensile test. SEM. 2800×

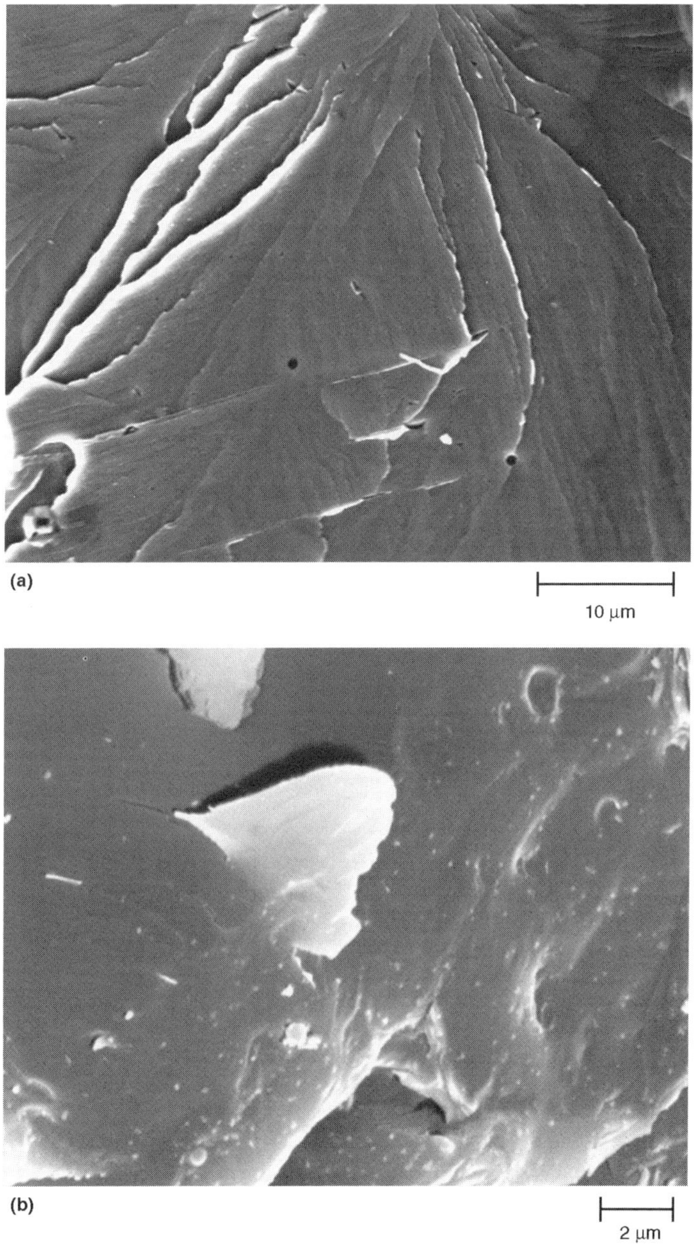

Fig. 2.5 Tongues on the transcrystalline cleavage fracture surface. (a) Ferrite grain in cast steel, impact test at −160 °C. SEM. 2000×. (b) Silicon in hypereutectic aluminum-silicon alloy, static tensile test. SEM. 5500×

Features of the fracture profile include (Fig. 2.11, 2.12):

- *Profile of the main crack:* the line of intersection of the macrosurface of fracture with the metallographic section plane
- *Profile of the secondary crack:* the profile of the branched crack, directly connected with main profile or formed in an isolated zone (internal crack)
- *Screen:* the region of the discontinuity (crack), on the profile line tilted under small angle to the macrosurface of the fracture, situated directly beneath it
- *Ligament:* the profile of the intersection of the neck, formed from α-aluminum solid solution, visible between two hard particles, for example, on the step profile (microneck, bridge) or in the region of the dendrite arm of the α-aluminum solid solution (macroligament) with the metallographic section plane
- *Step profile:* the line of the crack in two-phase region: on the parallel cleavage planes of the neighboring, brittle particles and in the soft matrix among them
- *Line of the shear ridge:* the line formed by cutting the shear region in the plastic phase with the metallographic section plane
- *Cleavage line:* the line formed by cutting of the crack surface in the brittle microstructure component with the metallographic section plane

2.2.5 Description of the Fracture Surface

Morphology of Cleavage Fracture Surfaces (Ref 1, 2, 5, 12, 15). Features of cleavage fracture surfaces include:

- *Cleavage facets* are the areas of the cleavage planes, characteristic for local crystallographic orientation and limited for the range of one grain; in case of the ideal cleavage fracture, they are smooth in the atomic scale (Fig. 2.3).
- *Cleavage steps* are the traces of the passage of the fracture front from one into another cleavage plane, usually parallel to the previous one (in range of one grain) (Fig. 2.4a, b).
- *Rivers and river patterns* are the system of the connected cleavage steps formed when low-angle, screw grain boundary, or screw dislocation are present on the fracture path. High-angle grain boundary or precipitates of the second phase are also obstacles for fracture propagation and can cause the formation of new river pattern. Rivers join along the crack propagation direction to minimize the surface energy. Absorption of the energy during fracture propagation, connected with increase in the separation surface area, causes a decrease in the level of brittleness (Fig. 2.4, 2.5a).
- *Tongues* are round steps, usually arranged in well-defined crystalline planes. Their formation can be connected with passage of the crack front by the region of the local plastic deformation, for example, deformation twins (Fig. 2.5).
- *Chevrons* are crossed steps, pointed out into direction of the local crack initiation point; they can be also the traces of the crossing of the local cleavage facet with twin system (Fig. 2.6).
- *Wallner lines* appear on the surface of the most brittle microstructure component, as a result of the interaction of the crack front with an elastic wave, the source of which is situated in another microregion of the material. The step system, formed in this way, is arranged in wave bands, which can intercross (contrary to the fatigue striations) (Fig. 2.7).

Morphology of Intercrystalline Fracture (Ref 1, 2, 5, 9 12, 15). Intercrystalline fracture can cross:

- Grain boundaries (Fig. 2.13a)
- Interdendritic regions (Fig. 2.13b)
- Interfaces (Fig. 2.13c)

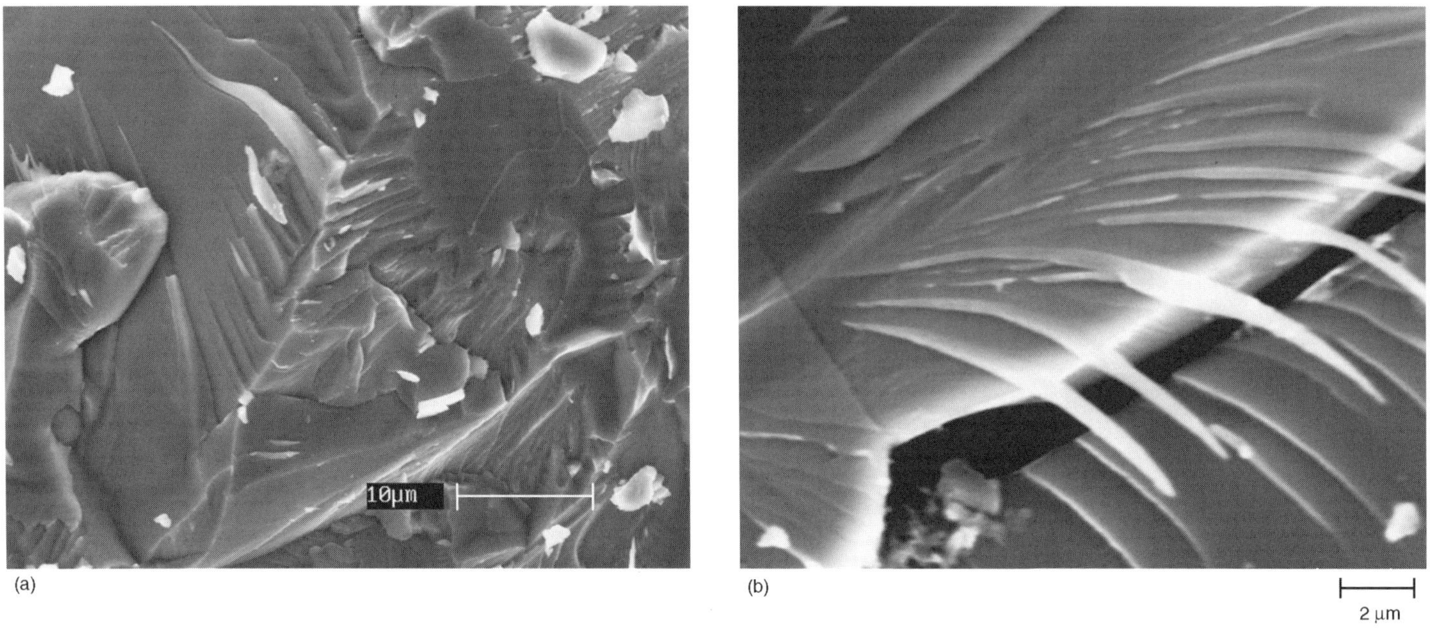

Fig. 2.6 Chevron formed by crossing steps, primary silicon in hypereutectic aluminum-silicon alloy, static tensile test. SEM. (a) 2000×. (b) 5500×

It reflects the morphology of these microstructure microregions.

Morphology of Plastic Fracture (Ref 1, 2, 5, 9, 10, 15). Features of plastic fracture surfaces include:

- *Pinpoint (reduction of area of the neck)* is the cracked microregion of the material (bridges, microligaments), previously plastically deformed, where nucleation and coalescence of microvoids did not occur (Fig. 2.14a, b).
- *Tear ridges,* in the microscale, are the line of the flow and cracking of the material in the region of the local neck. On both sides of the tear ridge, the plain fields are usually visible, characteristic of local decohesion by slip. Their presence reflects the discontinuity of the fracture process, caused by change of decohesion mechanism in neighboring microregions (Fig. 2.14c, d).

Morphology of Shear Fracture (Ref 2, 9, 11). Features of the shear fracture surface include:

- *Shear surfaces* are the plain regions of the shear of material, on the slip planes, favorably oriented from the viewpoint of slip-system activation. Usually they are observed in the microscale as an element of tear ridge in the grain (Fig. 2.14c, d) or in the dendrites of the plastic components of the material (Fig. 2.14e, f). In the macroscale, they are visible as the shear lips.
- *Shear dimples* are the effect of the shear process in the deformed microregions (see discussion below).

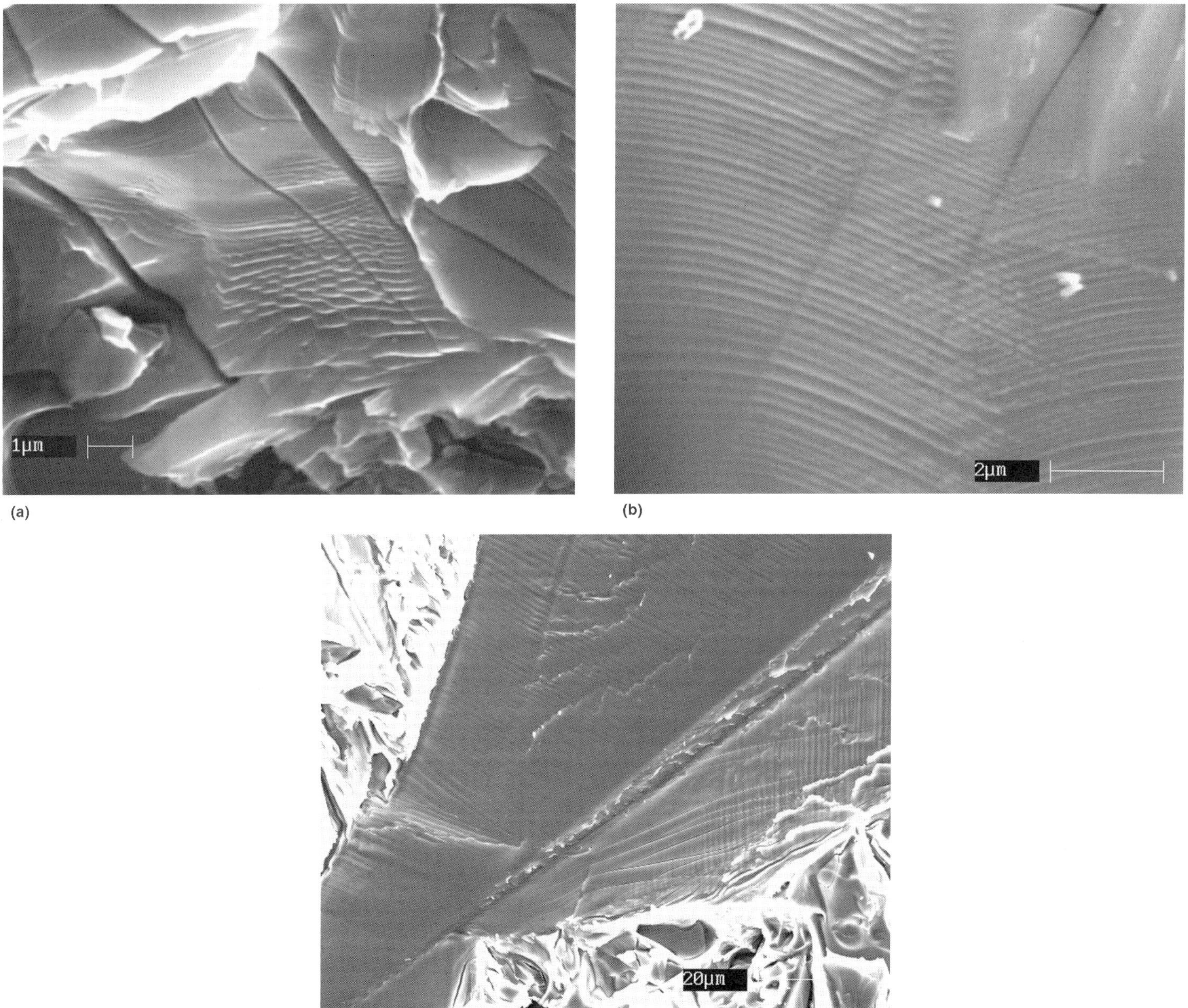

Fig. 2.7 System of wave steps (Wallner lines) on the surface of the cracked silicon precipitate in hypoeutectic aluminum-silicon alloy, impact test at −160 °C. SEM. (a) 8000×. (b) 10,500×. (c) 550×

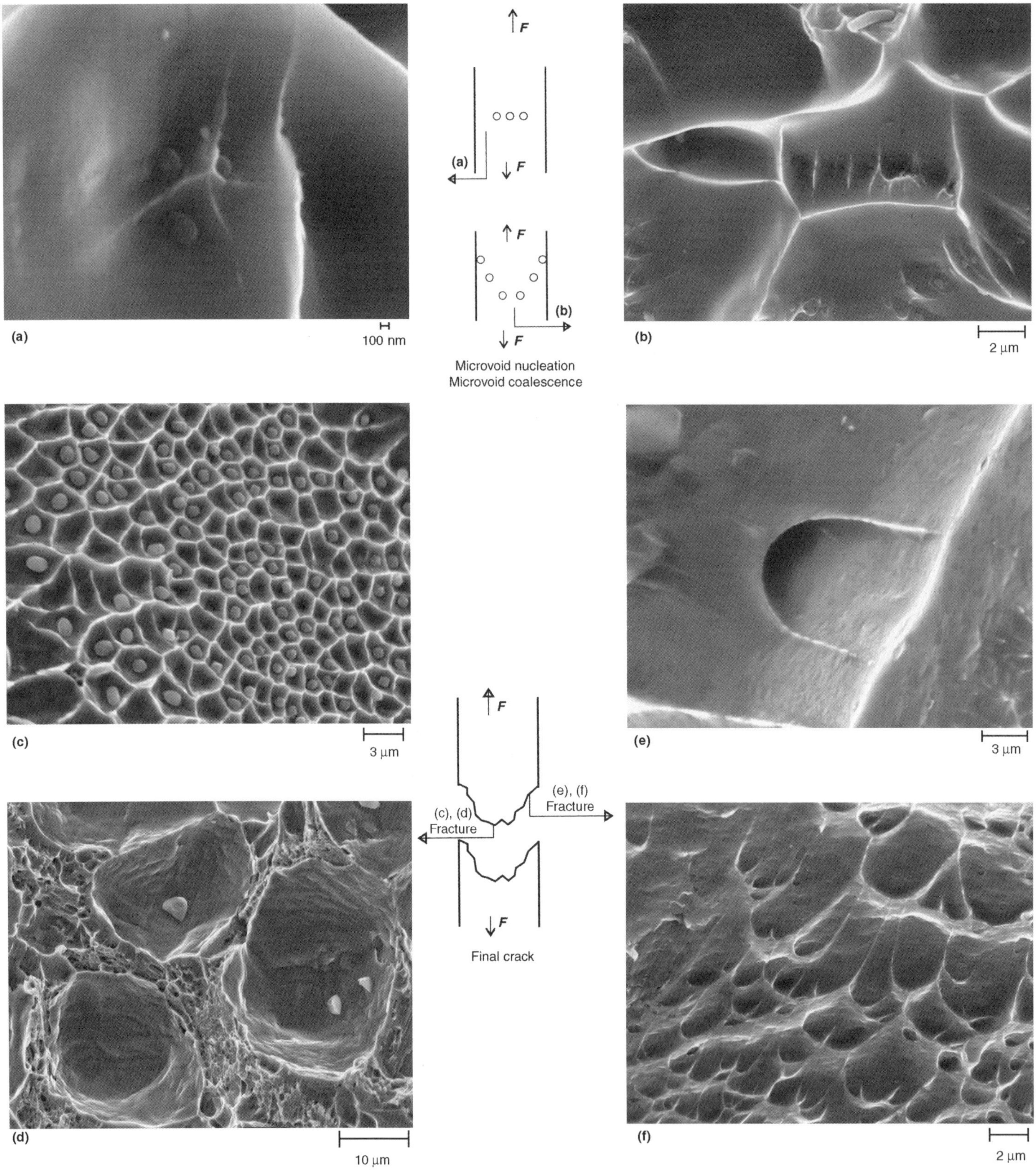

Fig. 2.8 Mechanism of formation of transcrystalline ductile fracture during static tensile test. ***F***, force.

Morphology of Ductile Fracture (Ref 1–3, 5, 9, 11, 12, 15). Features of the ductile fracture surface include:

- *Voids* are concave microregions of the initiation of the material decohesion, usually around the hard dispersed particles or other matrix discontinuities (Fig. 2.8a).
- *Dimples* are rounded hollows on the fracture surface. The shape and size of the dimples are determined by the size and distribution of the microstructure discontinuities (micropores, disperse particles, microcracks), plastic properties of the material, and the acting stresses. The dimples of different morphology can exist simultaneously on the surface of the ductile fracture, depending on the active local stress and strain states (Fig. 2.8c–f).
- *Equiaxial dimples* (ductile) form during uniaxial tensile (plain strain state) (Fig. 2.8c, e).
- *Tear dimples,* open or closed, form under complex stress state (e.g., tensile and bending or torsion). Round ends of the open tear dimples appear opposite the crack initiation region and are the same on both fracture surfaces.
- *Shear dimples* present in the shear areas of the plastically deformed material (Fig. 2.8f) are the oval hollows on the neck shear surface, in region of the plain stress state (Fig. 2.8d, f). Oval shear dimples formed during shear of the material can be open or closed. They are elongated into direction of the stress effect, and their coalescence takes place on the plane of the maximum shear stress.

The presence of dimples in the material proves that the plastic deformation takes place. Decohesion occurs on the successive parallel slip planes of favorable orientation. As a result of this process new, free surfaces are forming in material (Fig. 2.8, 2.14).

Morphology of Mixed Fracture (Ref 1, 2, 5). Mixed fracture can be characterized by simultaneous presence of the features of the cleavage and plastic fracture:

- Cleavage facets
- Steps
- Rivers, river patterns
- Tongues
- Tear ridges
- Dimples

Examples of the mixed fracture morphology are shown in Fig. 2.9 (brittle-ductile fracture) and in Fig. 2.10 (cellular fracture).

2.2.6 *Fatigue Fracture (Ref 1, 2, 12)*

Fatigue fractures belong to a particular group of fractures from the viewpoint of the formation mechanism and the material decohesion and the specific morphology of the surface.

Fatigue fracture can be classified according to the following criteria:

- The type of the loading
- The range on the Wohler's curve

According to this first criterion, the fractures can be defined as typical fatigue fractures or as fatigue fractures caused by:

- Thermal fatigue
- Corrosion
- Repeated impact loading
- Repeated loading of ultrasonic frequency

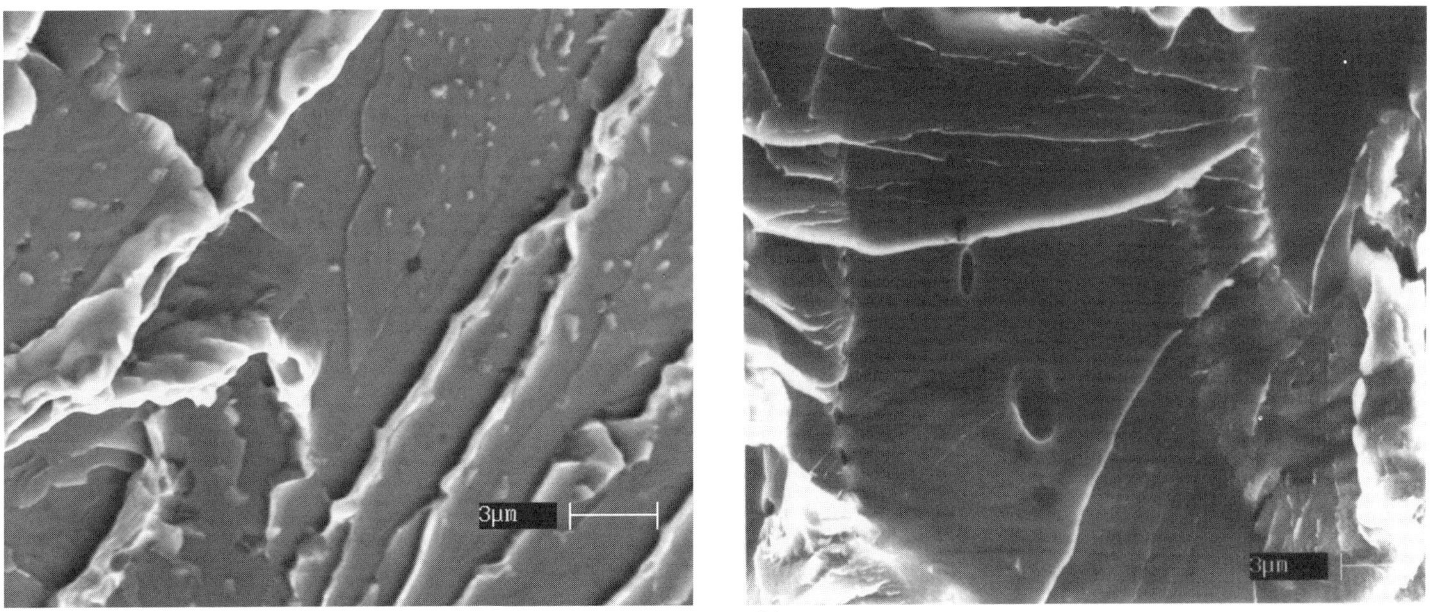

(a) (b)

Fig. 2.9 Mixed brittle-plastic fracture, high steps, dimples, tear ridge in ferrite grain in cast steel, impact test at room temperature. SEM. (a) 4700×. (b) 3500×

According to the second criterion, the following classifications can be defined:

- Fatigue fracture in the range of the short time and limited fatigue resistance
- Fatigue fracture in the range of the loading of the fatigue limit (characterized by presence of the plastic deformation)

Morphology of Fatigue Fracture. The characteristic features of fatigue fractures are fatigue striations (Ref 1, 2, 12) and the indent traces (Ref 1).

Fatigue striations are elongated bands of material, alternately concave and convex, parallel to the crack front. They are the traces of the crack propagation in each loading cycle and are situated perpendicularly to the crack propagation direction. In aluminum alloys, they are more continuous and regular than in steels. Brittle (Fig. 2.15a) and plastic striations (Fig. 2.15b) can be observed on fatigue fracture surfaces. The brittle striations are crossed with the perpendicular steps. The line of each step is parallel to the crack propagation direction. They are often present in dispersion-hardened aluminum alloys. In the macroscopic scale, fatigue lines are also visible on the fatigue fracture surface. In these regions, the

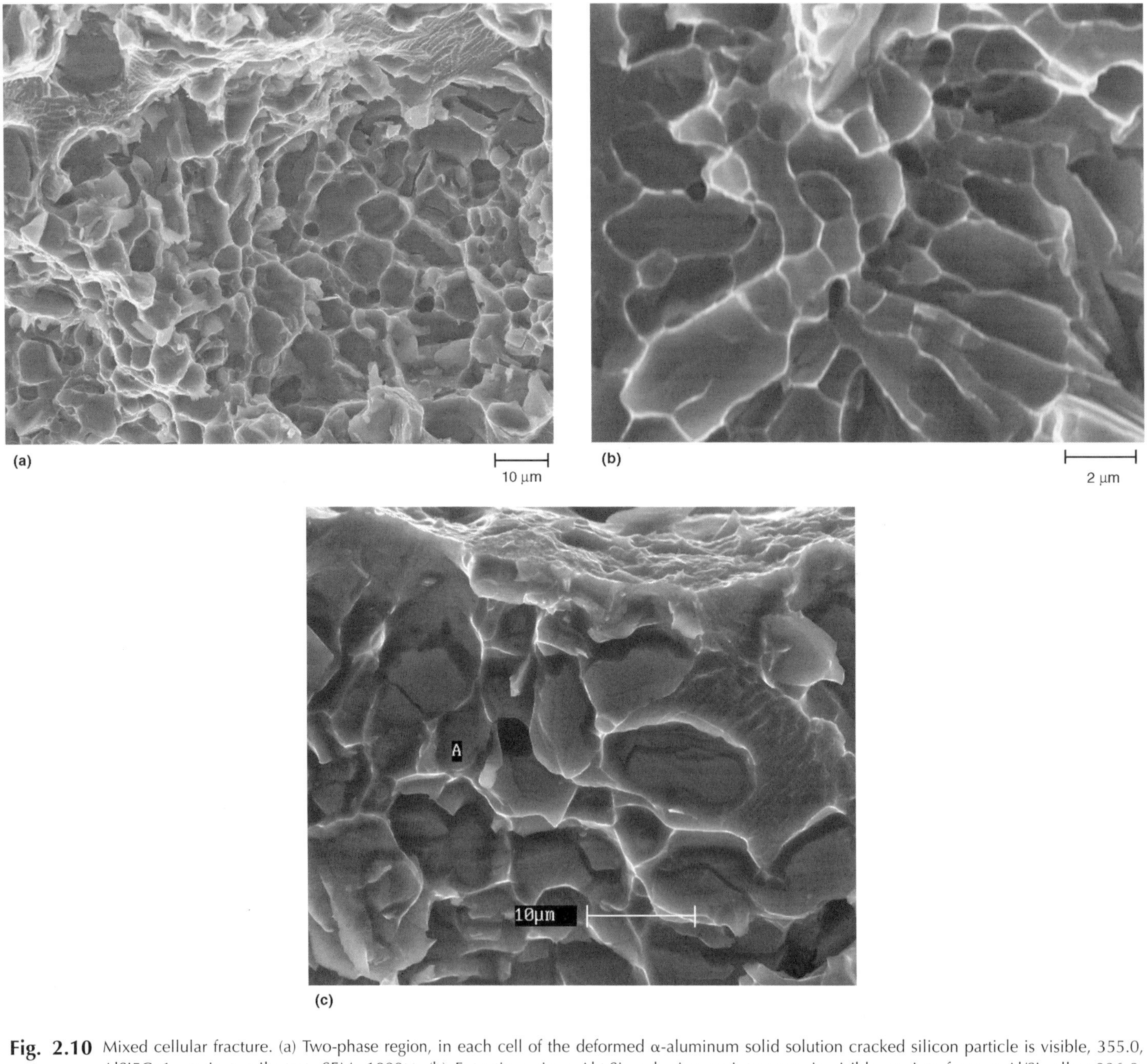

Fig. 2.10 Mixed cellular fracture. (a) Two-phase region, in each cell of the deformed α-aluminum solid solution cracked silicon particle is visible, 355.0, AlSi5Cu1, static tensile test. SEM, 1000×. (b) Eutectic grain α-Al+Si, cohesion maintenance is visible on interfaces α-Al/Si, alloy 336.0, AlSi13Mg1CuNi, modified, static tensile test. SEM. 6500×. (c) Cellular fracture with shear areas in matrix, cohesion maintenance is visible (at A) on interfaces α-Al/Si, alloy 355.0, AlSi5Cu, modified, static tensile test. SEM. 2000×

hardening of the material was stated (Ref 1). The distribution and spacing of the fatigue striations reflect the changes in the rate of the main crack propagation. Each fatigue line is composed of thousands of fatigue striations, so it represents several cycles of loading.

2.3 Quantitative Fractography

The main aim of quantitative fractography is to formulate a description of the fracture surface containing a numerical measure of the defined features of its morphology. It ought to verify:

- Numerical parameters of morphology description for complicated fracture surface and their measure
- Methods of the measurement of these parameters

In several works (Ref 4, 12, 16–21) concerning this problem, different solutions can be found, from the viewpoint of measurement methods and minuteness of detail.

To realize the aims of fractographic analysis mentioned previously—that is, to establish a statement of the relationships between mechanical properties and the mechanism destroying the material—it is necessary to evaluate:

- Degree of the development of the fracture surface (Fig. 2.16a)
- Fraction of the elements of defined morphology on the fracture surface (Fig. 2.16b)
- Quantitative description of the fracture surface features of defined morphology (Fig. 2.16c)

2.3.1 *Estimation of the Level of Surface Fracture Development*

The level of surface fracture development can be estimated by analyzing its profile line (Fig. 2.17). The simplest way is to analyze the fracture profile on the surface of the sample, but in this case, satisfying results can be obtained only for brittle fractures. Real, three-dimensional (3D) reconstruction for all kind of fractures will be possible when a sufficient number of profile sets have been captured. So, it is necessary to produce the sequence cuttings of the specimen. Each profile should be properly protected, revealed on the metallographic microsection, and subjected to a quantitative analysis. This method irreversibly destroys the specimen. Methods of nondestructive fracture analysis (Ref 2, 4, 21) include:

- Three-dimensional reconstruction and quantitative analysis on the base of the profile sequence, from fracture replicas, mounted in the resin contrasted according to specimen material
- Three-dimensional reconstruction and quantitative analysis of stereopairs (two coupled images, visible from different angles in the scanning electron microscope)
- Construction of the topographic maps using confocal laser microscopy or atomic force microscopy (enables direct measurement of the coefficient of surface development)

Table 2.2 lists selected fracture parameters and coefficients used for quantitative fracture analysis.

The coefficient R_s allows estimation of the value of the real fracture surface S, on the basis of its projection on the projection plane $A^{(n)}$:

$$S = R_s A^{(n)} \quad \text{(Eq 2.1)}$$

In Ref 17 it was proposed to estimate the real fracture surface coefficient R_s on the basis of the measurement of the value of the parameter R_l, estimated from the measurement results, carried out for the fracture profile, visible on the microsection:

$$R_s = f(R_l) \quad \text{(Eq 2.2)}$$

This approach is compatible with El-Soudani's rule (Ref 16–18), that two fractures of identical coefficients of the profile

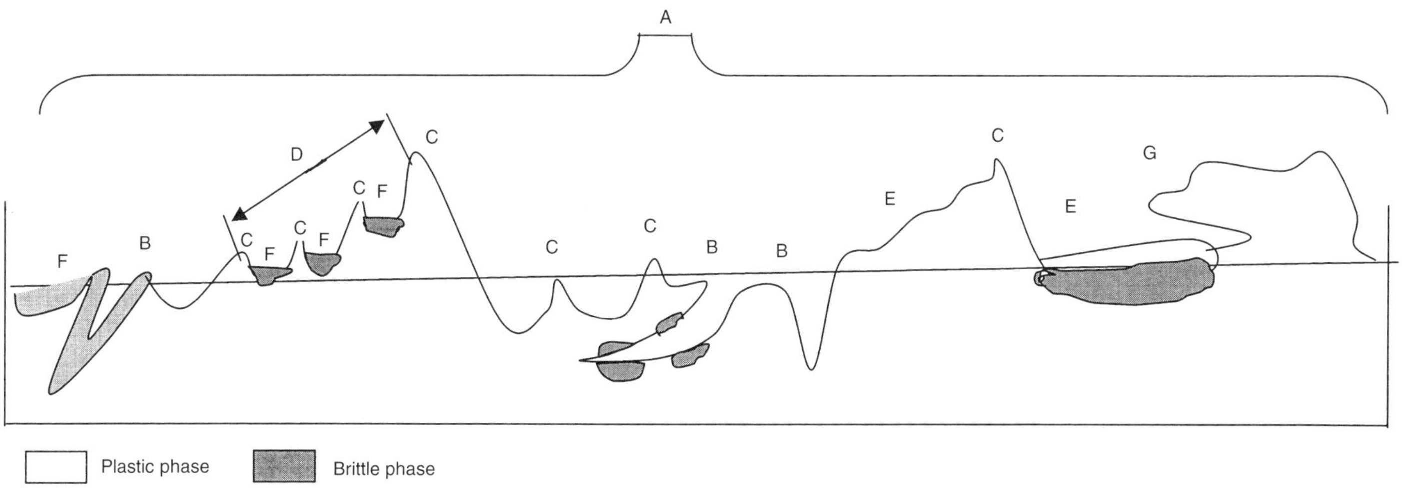

Fig. 2.11 Line of the fracture profile of polyphase alloy. A, profile of the main crack; B, profile of the secondary crack; C, fractured ligaments of the plastic deformed phase; D, step profile in two-phase region; E, line of shear in plastic phase; F, line of cleavage in brittle phase; G, screen

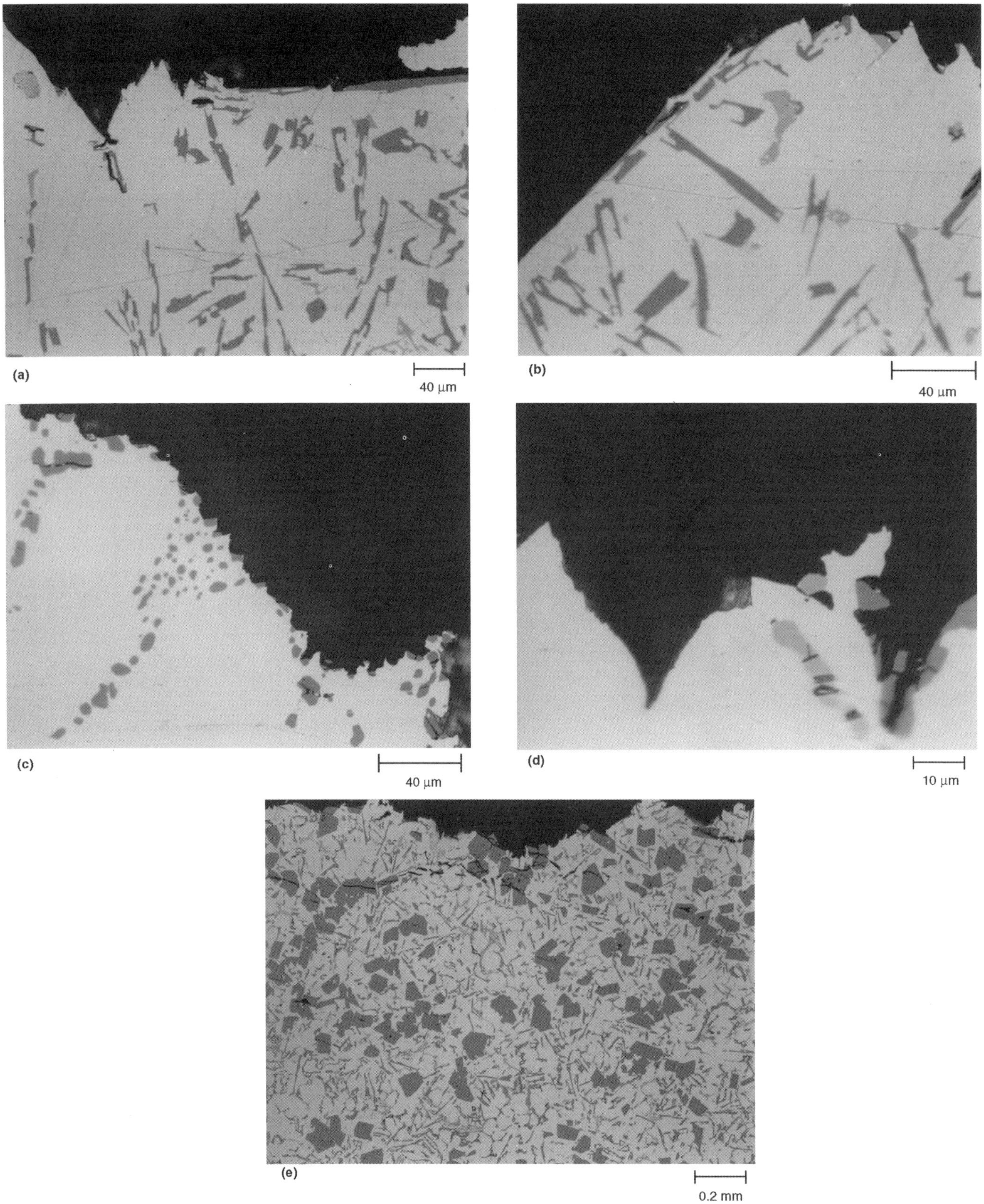

Fig. 2.12 Typical fracture profiles in aluminum-silicon alloys. (a) Line of shear, line of cleavage, fractured ligaments, secondary cracks, 359.0, AlSi9Mg, impact test at room temperature. 250×. (b) Line of shear, step profile, line of cleavage, 359.0, AlSi9Mg, impact test at room temperature. 400×. (c) Step profile, 356.0, AlSi7Mg, static tensile test. 400×. (d) Line of shear, fractured microligaments of the α-Al-solid solution, 356.0, AlSi7Mg, static tensile test. 1000×. (e) Line of shear, cleavage line, fractured bridges of the α-Al solid solution plastic, secondary cracks, 390.0, AlSi21CuNi, static tensile test. 50×

development R_l have the identical coefficients of the surface development R_s. Relationships between these factors is given by the formula:

$$R_s = \frac{2}{\pi - 2}(R_l - 1) + 1 \quad \text{(Eq 2.3)}$$

The relationships between real fracture surface S and coefficient of the profile development R_l, given by Eq 2.4, was verified experimentally:

$$S = (1.75\, R_l - 0.75)\, A^{(n)} \quad \text{(Eq 2.4)}$$

The fractal dimension D can also be used for estimation of the fracture surface development. A schematic of the synthetic fractal structure, representing the rough surface considered the model of the fracture surface, is shown in Fig. 2.18.

Fractal dimension for this structure is calculated with:

$$D = \frac{(\log N)}{(\log 1/n)} \quad \text{(Eq 2.5)}$$

where N is the number of the segment of the initial fractal motive, $1/n$ is the number of the partition of initial line, forming the individual fractal motive, and L_0 (Fig. 2.18) is the length of the initial element, without fractal motive.

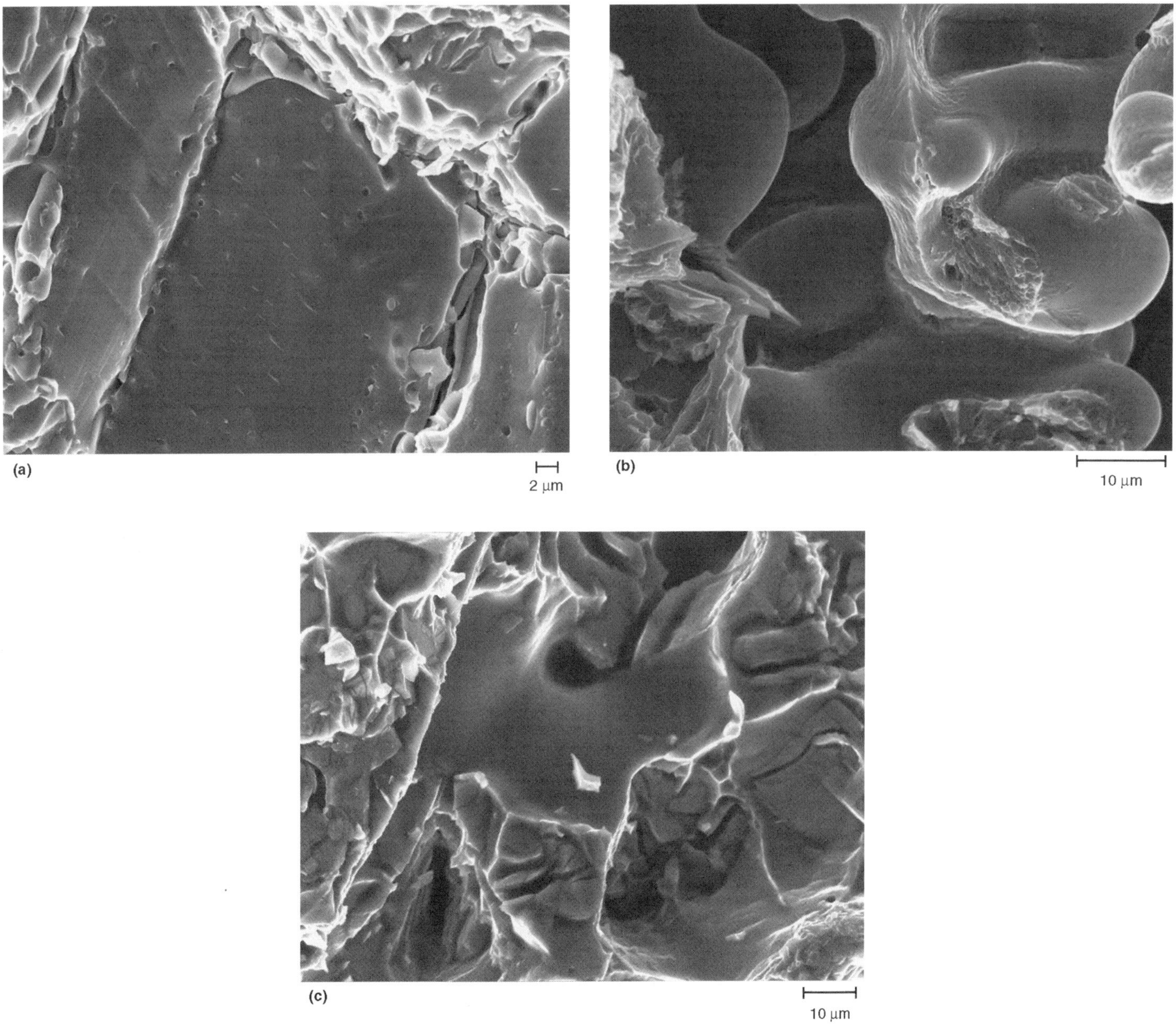

Fig. 2.13 Intercrystalline fracture. (a) Fracture on the grain boundaries, alloy 7075, static tensile test. SEM. 2000×. (b) Interdendritic fracture, hypoeutectic aluminum-silicon alloy, static tensile test. SEM. 1600×. (c) Fracture on interface, hypoeutectic aluminum-silicon alloy, static tensile test. SEM. 1000×

The relationship between the coefficient of the profile development and the measurement step is used to estimate the fractal dimension *D* of the fracture profile. It can be measured according to Mandelbrodt's or Minkowski's scheme (Ref 17–19). Image transformation using mathematical morphology methods is very often carried out before the measurements.

During the investigation of the sintered carbides, the relationship between all the crack energy during static bending and the coefficient of the profile development R_l was stated (Ref 21).

2.3.2 Estimation of the Surface Fraction of the Microregions for Fractures of Defined Morphology

After estimation of the real fracture surface *S* or real profile line length, the values of the parameters shown in Fig. 2.16(b) can be measured by means of the line method (Ref 4, 18). In this manner, the surface fraction of the fracture microregions of defined

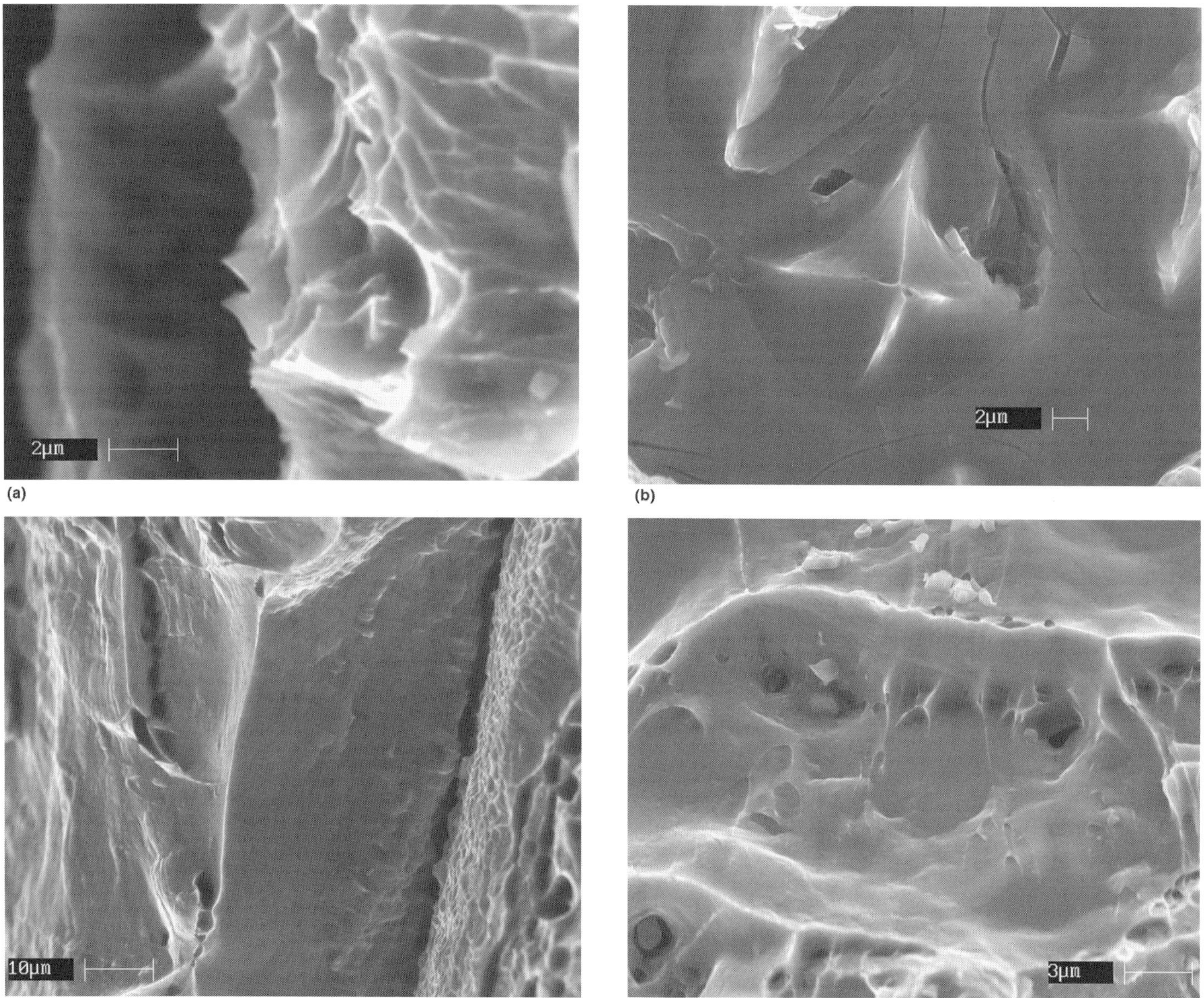

(a) (b) (c) (d)

Fig. 2.14 Characteristic features of surface morphology of transcrystalline, plastic and ductile fracture. (a) Fractured microneck in deformed α-Al solid solution, visible point reduction of area, hypoeutectic aluminum-silicon alloy, static tensile test. SEM. 6500×. (b) Fractured microneck, shear voids, hypoeutectic aluminum-silicon alloy, static tensile test. SEM. 3200×. (c) Tear ridge in the α-aluminum solid solution, static tensile test, alloy 7075. SEM. 4000×. (d) Tear ridge in the α-aluminum solid solution, visible traces of the deformation in form of the shear dimples, static tensile test, alloy 7075. SEM. 4000×. (e) Local shear surfaces in α-Al solid solution, hypoeutectic aluminum-silicon alloy, static tensile test. SEM. 900×. (f) Local shear surfaces in α-aluminum solid solution, oval shear dimples, hypoeutectic aluminum-silicon alloy, static tensile test. SEM. 6000×. (g) Equiaxial and oval shear voids around disperse particles of the $MgZn_2$ phase, decohesion on the interfaces, alloy 7075, static tensile test. SEM. 15,000×. (h) Shear oval dimples formed after coalescence of the linear void sequence, alloy 7075, static tensile test. SEM. 2500×. (i) Area of the ductile fracture with small equiaxial dimples, rounded with band of the shear dimples initiated on the intermetallic inclusions, cast steel, impact test. SEM. 6000×

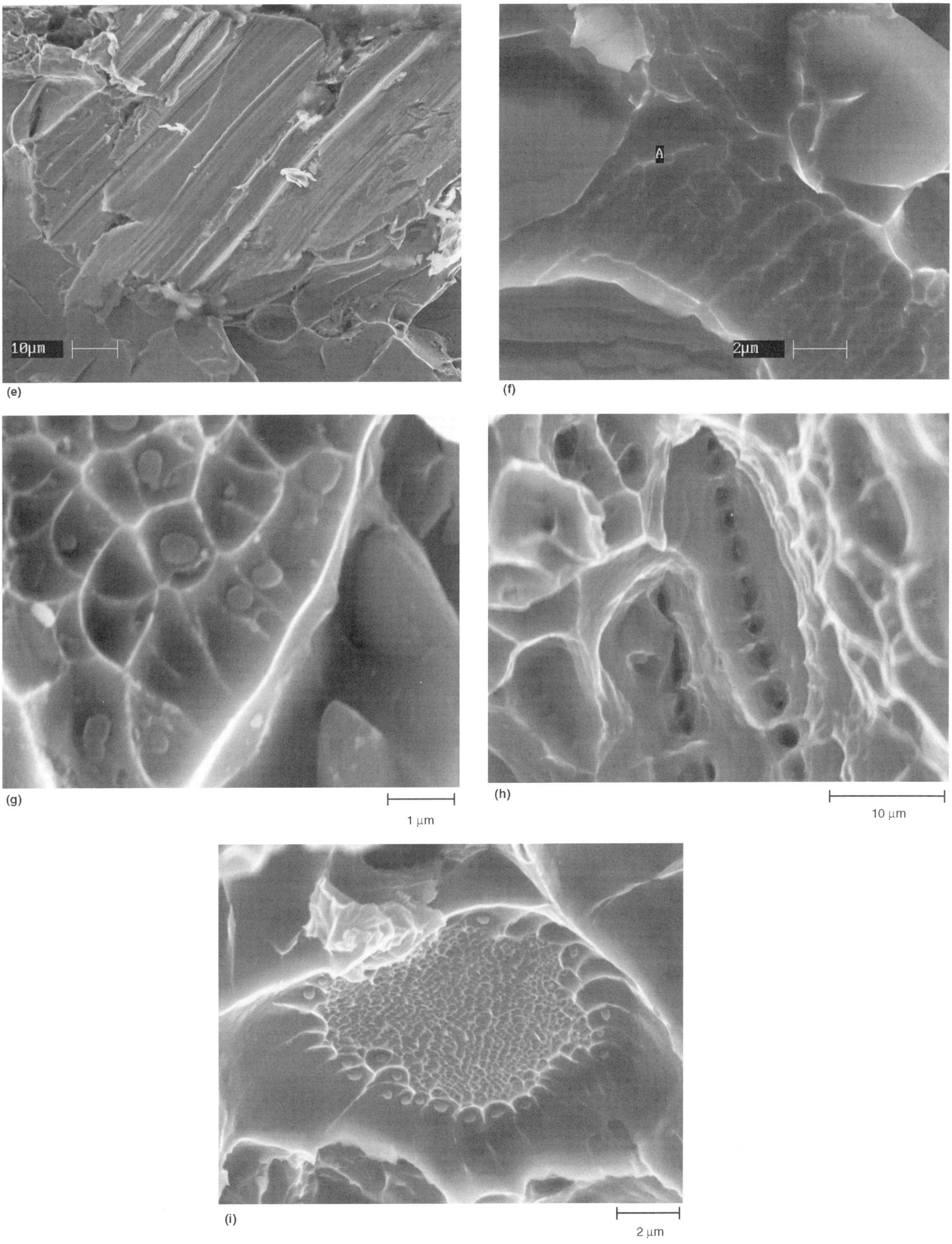

Fig. 2.14 (continued)

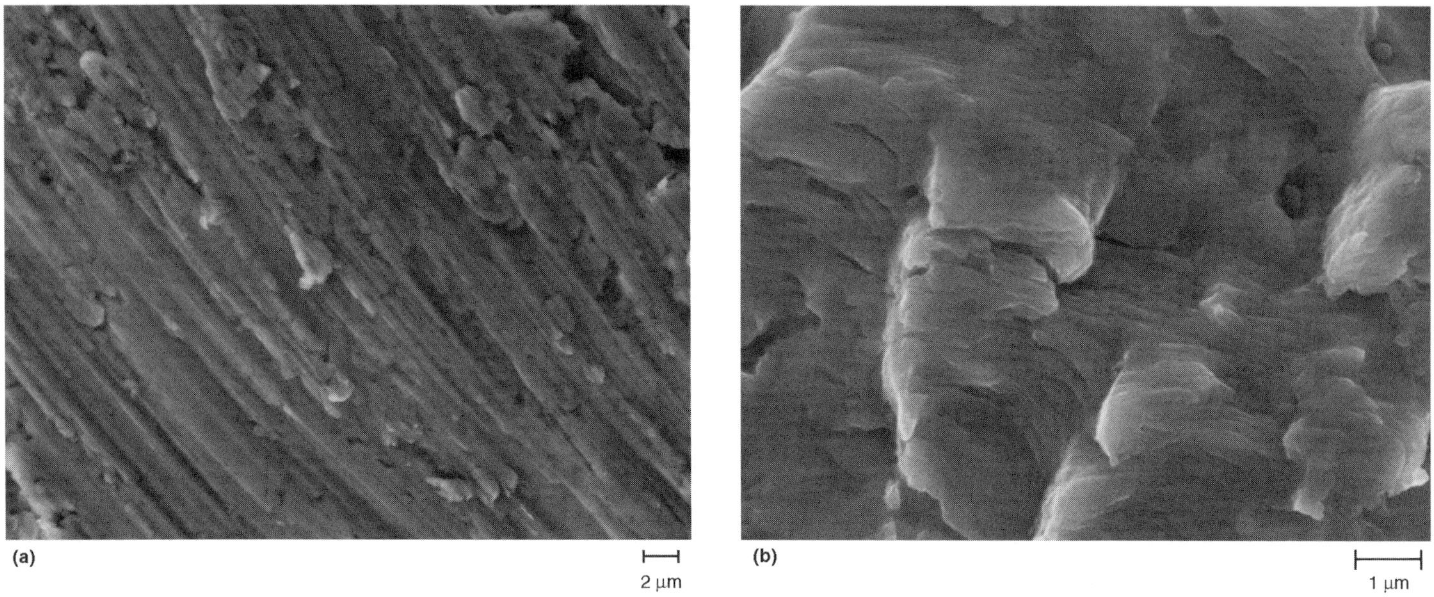

Fig. 2.15 Fatigue fracture. (a) Brittle fatigue striations, alloy 355.0, AlSi5Cu. SEM. 5000×. (b) Plastic fatigue striations, cast steel. SEM. 15,000×

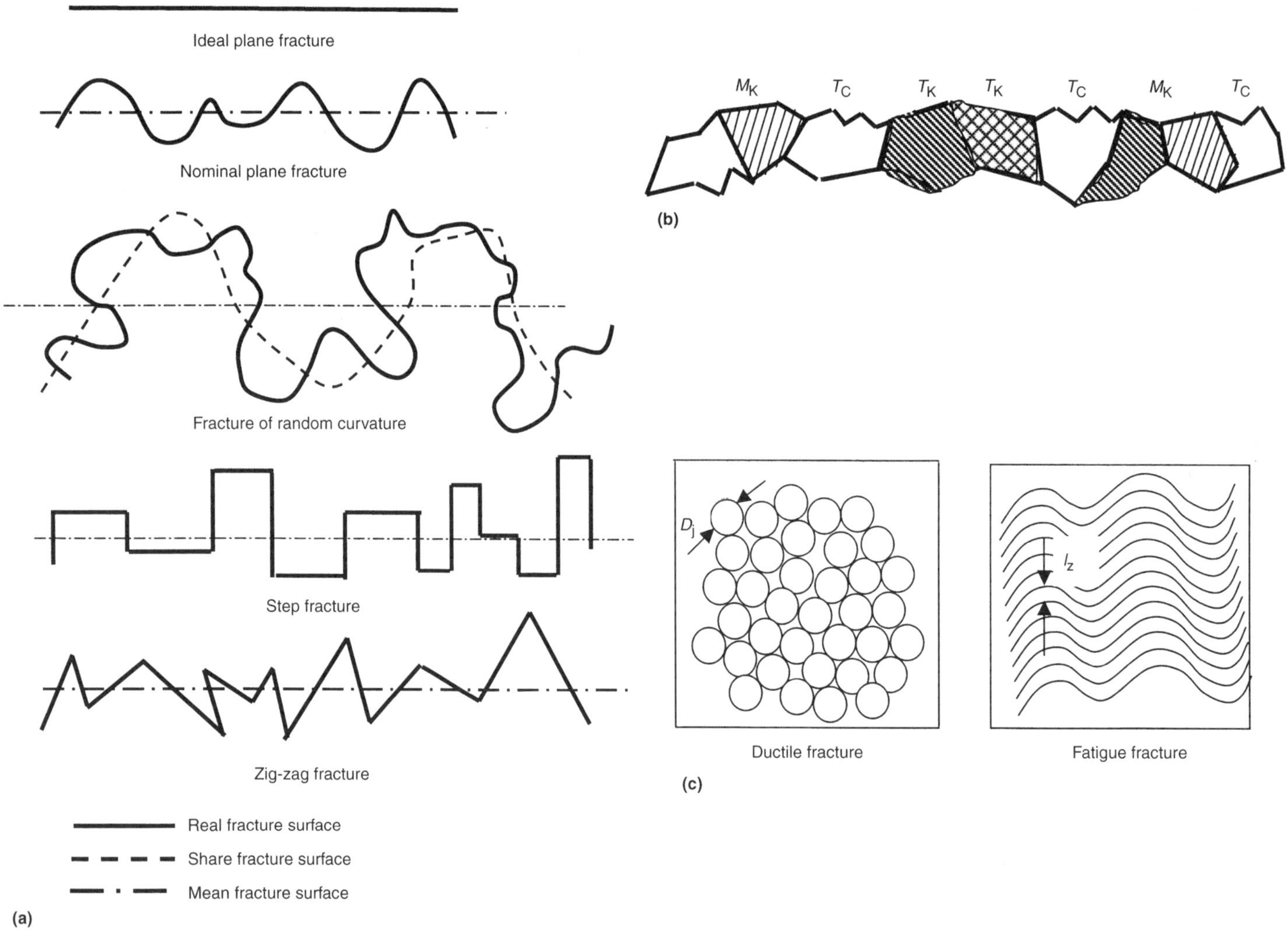

Fig. 2.16 Quantitative fracture characterization. (a) Classification of fractures according to their morphology and development of surface. Source: Ref 4, 16, 18. (b) Quantitative characteristics of mixed fracture surface. M_K, elements of intercrystalline brittle fracture; T_K, elements of brittle transcrystalline fracture; T_C, elements of ductile transcrystalline fracture. Source: Ref 4, 18. (c) Quantitative characteristics of revealed fracture features. D_j, dimple diameter; l_z, striation distance

morphology can be estimated, and information about crack energy and its local mechanism can be obtained.

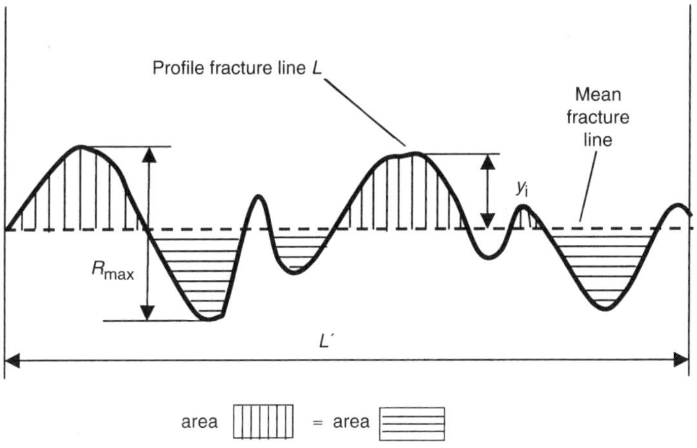

Fig. 2.17 Description of the fracture surface development, quantitative parameters estimated from profile line. L, true length of the profile line; L', projection of the profile line on the projection plane. Source: Ref 4, 16, 18

Table 2.2 Selected parameters describing the fracture surface

No.	Parameter	Designation	Definition
1	Mean arithmetical deviation of the profile from the average profile line	R_a	$R_a = 1/n\Sigma y_i$
2	Maximum height of irregularity	R_{max}	$R_{max} = y_{max} - y_{min}$
3	Line factor of the profile development	R_l	$R_l = L'/L$
4	Wave factor	P_s	$P_s = R_l$
5	Mean curvature of convex profile elements	K^+	. . .
6	Mean curvature of concave profile elements	K^-	. . .
7	Average curvature	K_{sr}	$K_{sr} = (K^+ + K^-)$
8	Factor of the development of the fracture surface	R_s	$R_s = f(R_l)$
9	Fractal dimension	D	$D = \log N/\log(1/n)$

See Fig 2.17 for definition of L', L, and y_i. Source: Ref 2, 4, 16–20

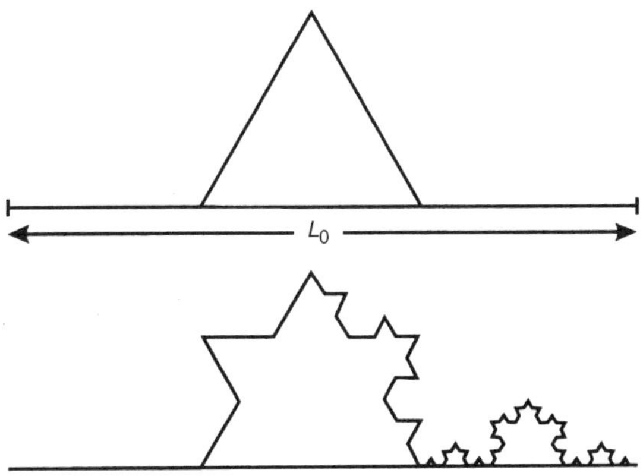

Fig. 2.18 Synthetic fractal structure. Source: Ref 19, 20

2.3.3 Quantitative Characteristics of the Fracture Areas of Defined Morphology

Estimation and measurements of the defined features of the fracture morphology make it possible to find a direct correlation among coefficients describing the fracture morphology (Fig. 2.16c) and its mechanism and the mechanical properties of the material. For example, an empirical relationship between fatigue striations distance l_z and stress-intensity factor ΔK can be used (Ref 1, 12):

$$l_z = 6(\Delta K/E)^2 \quad \text{(Eq 2.6)}$$

where E is elastic modulus.

Other examples of the relation between some characteristic feature of the fatigue fracture morphology and material properties are shown in Ref 1. The value of l_z (distance of the fatigue striations) was used for calculating the crack energy and the crack path reconstruction during fatigue fracture.

The authors of Ref 22 have evaluated the relationship between a stress state (triaxiality factor) in material and the mean area and the mean diameter of the dimples visible on the fracture surface (Fig. 2.16c). The analysis has concerned the ductile fracture of the cast steel under triaxial stress state.

REFERENCES

1. S. Kocańda, *Zmęczeniowe pękanie metali (Fatigue Failure of Metals),* Wyd. Naukowo-Techniczne, Warsaw, 1985 (in Polish)
2. *Fractography and Atlas of Fractographs,* Vol 9, 8th ed., *Metals Handbook,* American Society for Metals, 1974
3. M. Warmuzek, Zastosowanie analizy fraktograficznej do określenia wpływu stanu strukturalnego na wybrane własności stopów odlewniczych (Fractography as a Tool of Analysis of the Influence of the Structure State on the Properties of the Cast Alloys), *Praca Statutowa IO,* No. 5020, 1997 (in Polish)
4. J. Cwajna, A. Maciejny, and J. Szala, Aktualny stan i kierunki rozwoju fraktografii ilościowej (State and Prospects of Development of the Quantitative Fractography), *Inż. Materiałowa,* Vol 5 (No. 6), 1984, p 161–176 (in Polish)
5. M. Richard, J.C. Mercier, and S. Jacob, La microfractographie des alliages d'aluminium moules, *Fonderie Fondeur d'Aujourd'hui,* Vol 36, 1984, p 13–19 (in French)
6. A. Hałas and H. Szymański, *Mikroskopy elektronowe (Electron Microscopes),* Wyd. Komunikacji i Łączności, Warsaw, 1965 (in Polish)
7. G. Schimmel, *Metodyka mikroskopii elektronowej (Experimental Methods for Electron Microscopy),* PWN, Warsaw, 1976 (in Polish)
8. M. Warmuzek, Zastosowanie mikroskopu skaningowego w badaniach materiałoznawczych (Scanning Electron Microscope as a Tool in Material Testing), *Przegl. Lit., Praca IO,* No. 2970, 1987 (in Polish)
9. M.F. Ashby, C. Ghandi, and M.R. Taplin, Fracture-Mechanism Maps and Their Construction for FCC Metals and Alloys, *Acta Metall.,* Vol 27, 1979, p 699–729

10. G.E. Dieter, *Mechanical Metallurgy,* 3rd ed., McGraw-Hill, 1986, p 241–271
11. M. Biel-Gołaska, Analysis of Cast Steel Fracture Mechanism for Different States of Stress, *Fatigue Fract. Eng. Mater. Struct.,* Vol 21, 1998, p 965–975
12. R.W. Hertzberg, *Deformation and Fracture Mechanics of Engineering Materials,* John Wiley & Sons, 1989
13. M.L. Bernsztejn and W.A. Zajmowskij, *Struktura i własności mechaniczne metali (Structure and Mechanical Properties of Metals),* Wyd. Naukowo-Techniczne, Warsaw, 1973 (in Polish)
14. D.R. Askeland, *The Science and Engineering of Materials,* PWS-Kent Publishing Co., 1987
15. P. Reznịcek and J. Stetina, Fractography of an Al-Si12-Cu-Mg-Ni Cast Alloy, *Prakt. Metallogr.,* Vol 16 (No. 2) Feb 1979, p 59–66
16. S.M. El-Soudani, Theoretical Basis for the Quantitative Analysis of Fracture Surfaces, *Metallography,* Vol 7, 1974, p 271–311
17. M. Coster and J.L. Chermant, Recent Developments in Quantitative Fractography, *Int. Met. Rev.,* Vol 28, 1983, p 228–249
18. L. Wojnar, 10 lat rozwoju fraktografii ilosciowej (1983–1993), *Inż. Materiałowa,* No. 4, 1993, p 89–99 (in Polish)
19. E.E. Underwood and K. Banerji, Fractals in Fractography, *Mater. Sci. Eng.,* Vol 80 (No. 1), June 1986, p 1–14
20. N. Jost and E. Hornbogen, On Fractal Aspects of Metallic Microstructures, *Prakt. Metallogr.,* Vol 25 (No. 4), April 1988, p 157–173
21. S. Roskosz, Porównanie metod ilościowego opisu przełomów węglików spiekanych (Comparison of the Quantitative Methods for Description of the Fracture of the Sintered Carbides), *Wiad. Stereologiczne,* 1998, p 12–20 (in Polish)
22. M. Biel-Gołaska and L. Gołaski, The Analysis of the Ductile Failure Process of Cast Steel Subjected to Triaxial Stress States, *Prace IO,* Vol 44 (No. 1–2), 1994, p 38–57

CHAPTER 3

Microstructural Aspects of the Failure of Aluminum-Silicon Casting Alloys

MORPHOLOGICAL ANALYSIS of the fracture surface shows the importance of alloy microstructure in the cracking process. Decohesion mechanism and crack path are strongly influenced by the volume fraction and morphology of the silicon precipitates (Ref 1–3). Due to these factors, two general fracture modes of these alloys can be identified:

- Transcrystalline brittle fracture (mainly alloys in as-cast state)
- Cellular fracture (modified or heat treated alloys)

3.1 Transcrystalline Brittle Fracture

In alloys with significant silicon volume fraction ($\geqslant$ 1.65 wt% Si), where the eutectic silicon forms a continuous network, the crack propagates on the silicon cleavage planes or other brittle microstructure components as different intermetallic phases (see Fig. 6.13, 6.27, 6.36, and 7.51). The sharp edges or ends of the brittle particles are preferred crack initiation sites (see Fig. 7.56 and 7.59). Energy is consumed for forming two new surfaces and to overcome the work on the cleavage planes. An increase of the fracture surface due to specific fracture features (steps, river patterns, tongues) results in some increase in the fracture work (see Fig. 6.25, 8.21, 8.24, and 9.7). The commonly observed connections of the cleavage steps reflect the tendency to minimize the fracture work along the crack path. The very specific feature of the fracture of hypereutectic alloys in primary silicon precipitates is Wallner lines (see Fig. 8.18). Very rarely, the slip trace also can be observed in these precipitates as a result of the local slip trace (of planar glide type) system activation (see Fig. 8.35). In these alloys, α-aluminum solid solution is slightly deformed in small areas visible as necks or ligaments on the fracture profile line (see Fig. 6.32, 7.54, and 8.19) or tear ridges in the fracture surface.

3.2 Cellular Fracture

Fracture of cellular morphology forms in most cases in alloys of modified silicon morphology (in liquid or solid state). The silicon precipitates become rounded or fibrous; the eutectic network is partially broken. The brittle particles are surrounded with a relatively soft matrix and sometimes isolated. Due to the strong cohesion at the interfaces between α-aluminum and silicon, the matrix is deformed under local active stress (see Fig. 4.16, 4.21, and 5.26). The cells are formed around the silicon-cracked particles by plastic deformation of the matrix. The traces of these events are visible as necks or ligaments on the fracture profile or the high tear ridges on the fracture surface. In the alloys after heat treatment of T6 type (dispersion strengthening), the typical ductile fracture mechanism can take place (see Fig. 5.31) in the matrix. The mechanism of local decohesion, initiated at α-aluminum/brittle particle interfaces, is very often composed: voids and dimples coalesce on the cell ridges, on both sides of the tear line. The shear surfaces, plane or with open shear dimples, are formed as a result of the successive slips in the numerous near slip planes in the aluminum crystal lattice. Decohesion at the interfaces between α-aluminum and silicon is rather rare, while the secondary cracks in the brittle phases can be observed very often.

A summary of failure mechanisms of the cast aluminum-silicon alloys is presented in Table 3.1.

Table 3.1 General failure mechanisms in cast aluminum-silicon alloys

Microstructure	Fracture morphology	Crack initiation	Crack path	Fracture mechanism
Brittle-phase continuous network and compact massive precipitates of sharp edges in soft matrix	Transcrystalline, brittle	Brittle-phase precipitates	Cleavage planes of brittle-phase precipitates	Cleavage in brittle phase
Broken network of brittle-phase, rounded isolated precipitates in soft matrix	Transcrystalline, cellular (brittle and plastic or ductile)	Brittle-phase precipitates	Cleavage planes of brittle-phase and deformed plastic matrix	Cleavage in brittle phase; slip in soft matrix

REFERENCES

1. M. Warmuzek and K. Rabczak, Microscopic Analysis of the Microstructure Aspects of Multiphase Al-Si Alloy Failure, *Proc. of 7th European Conference on Advanced Materials and Processes*, presented at Euromat 2001 (Rimini, Italy), 10–14 June 2001
2. P. Reznicek and J. Stetina, Fractography of an Al-Si12-Cu-Mg-Ni Cast Alloy, *Prakt. Met.,* Vol 16 (No. 1), 1979, p 59–66
3. M. Richard, J.C. Mercier, S. Jacob, La Microfractographie des Alliages d`Aluminium Moules (Fractography of the Aluminum Casting Alloys), *Fonderie Fondeur d'Aujourd'hui,* Vol 36, 1984, p 13–19

CHAPTER 4

Alloy 336.0 (AlSi13Mg1CuNi)

Composition and Properties

Alloy	Designation	Chemical composition (main components), wt%						Properties(a)	
		Si	Cu	Mg	Mn	Ni	Fe	R_m, MPa	A_5, %
As-examined	AlSi13Mg1CuNi	11.5–13.0	0.8–1.5	0.8–1.5	...	0.8–1.3	0.8	220	0.5
Aluminum Association standard (b)	336.0	11.0–13.0	0.50–1.5	0.7–1.3	0.35 max	2.0–3.0	1.2 max	214	0.5

(a) R_m, ultimate tensile strength; A_5, elongation measured over a length of 5.65 S_o, where S_o is the cross-sectional area of the test specimen before the test. (b) Alloy 336.0 is registered with the Aluminum Association and designated by ASTM as a permanent mold casting alloy. R_m is the minimum ultimate tensile strength for permanent mold 336.0-T551.

Microstructures

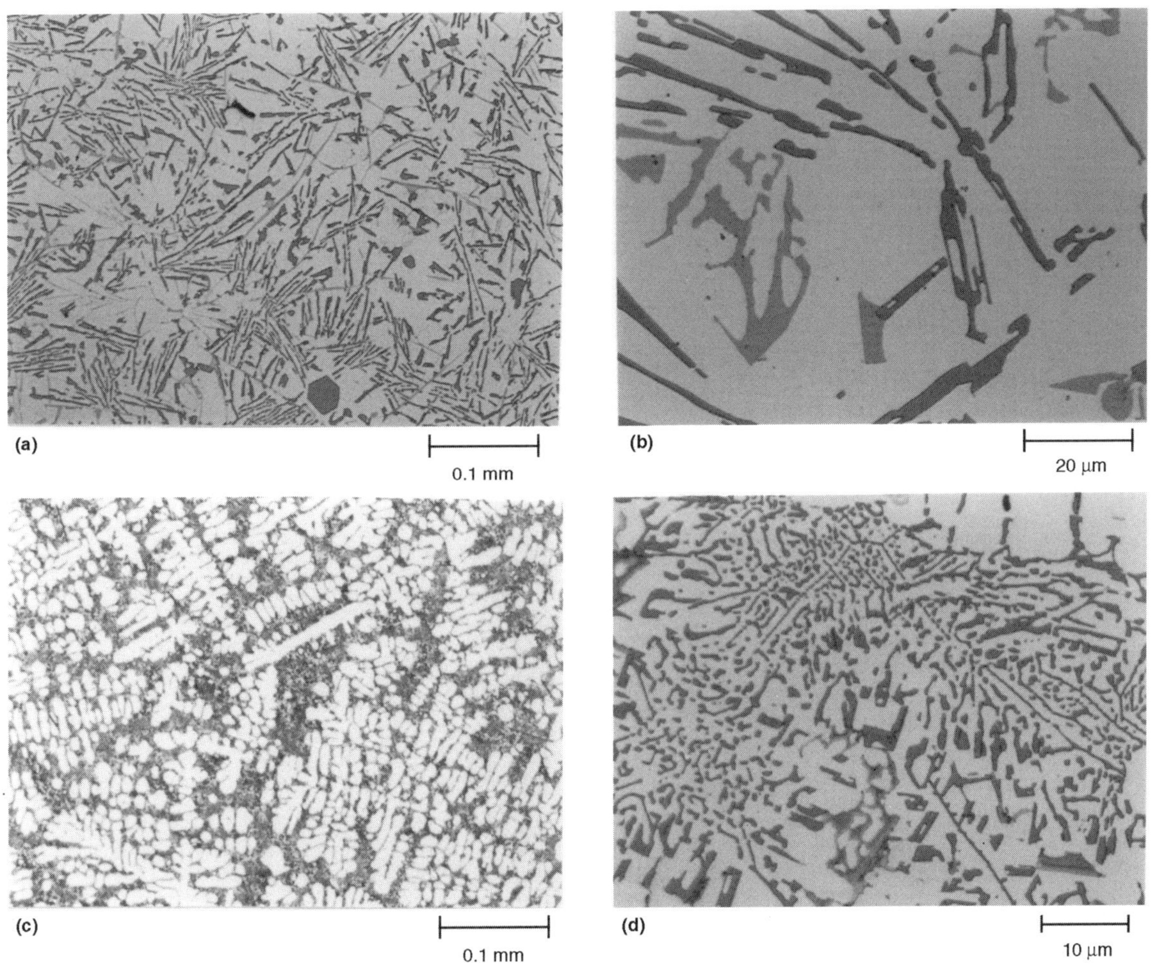

Fig. 4.1 Microstructures of AlSi13Mg1CuNi (Alloy 336.0). Light microscope micrographs; etched with 1% HF. (a) As-cast (F). 150×. (b) As-cast (F). 750×. (c) As-cast modified. 150×. (d) As-cast modified, 1200×

Fracture Profiles of Alloy 336.0 (AlSi13Mg1CuNi), Refined, Modified, Die Cast Parts

Fig. 4.2 Fracture profile of specimen after static tensile test. The main crack crossed the boundary zone: eutectic/dendrites of the α-aluminum solid solution. 50×

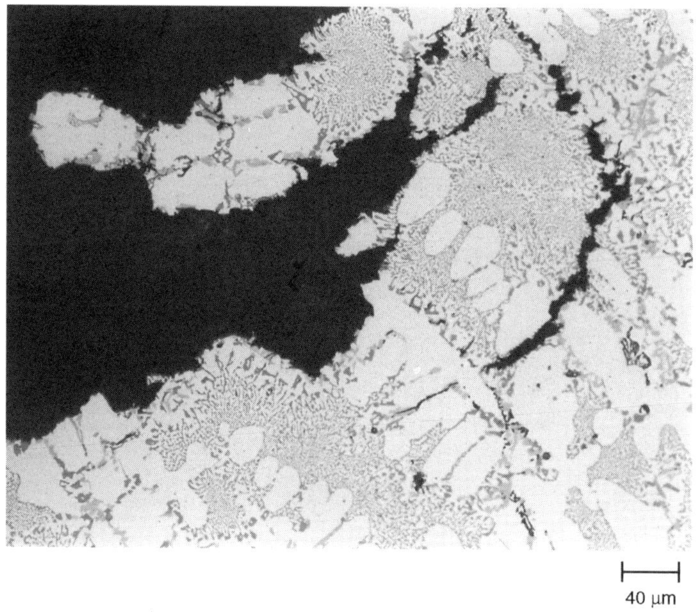

Fig. 4.3 Fracture profile of specimen after static tensile test. The main crack crossed the boundary zone: eutectic/dendrites of the α-aluminum solid solution. The screen, formed in this region, is shown. The sharply ended ligaments of the dendrite arms of α-aluminum solid solution are visible among cracked particles of the eutectic silicon. The shear edges of the dendrites of the α-aluminum solid solution and the secondary cracks in two-phase regions also can be observed. 200×

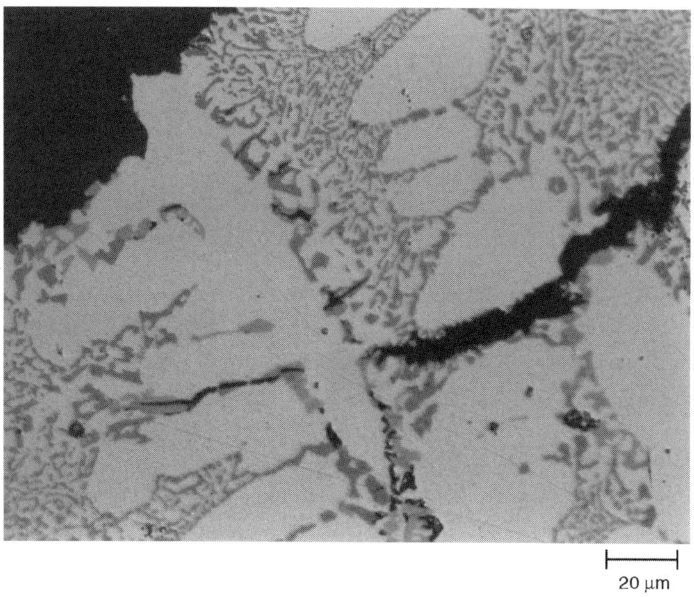

Fig. 4.4 Detail of the profile visible in Fig. 4.3. The microligaments of the α-aluminum solid solution, situated among the brittle eutectic phases, have cracked. In two-phase regions, secondary cracks have formed. 500×

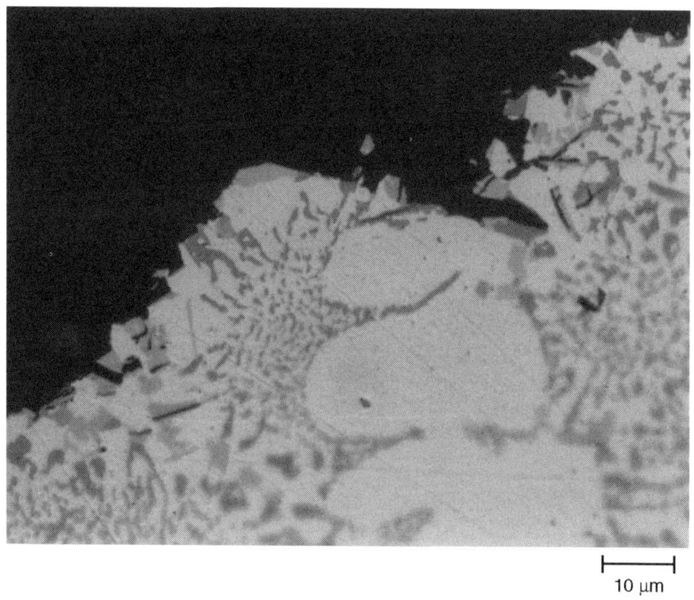

Fig. 4.5 Fracture profile of specimen after static tensile test. The main crack crossed the eutectic region and propagated by the parallel cleavage planes in precipitates of the brittle eutectic phases, silicon and Al_6Cu_3Ni. Secondary and internal cracks in brittle eutectic phases are visible. 1000×

Fracture Surfaces for Alloy 336.0 (AlSi13Mg1CuNi), Refined, Metal Mold Cast Part, Fracture after Static Tensile Test

Note: Alloy 336.0 is registered with the Aluminum Association and ASTM for use in permanent mold castings. See the table for the differences in composition.

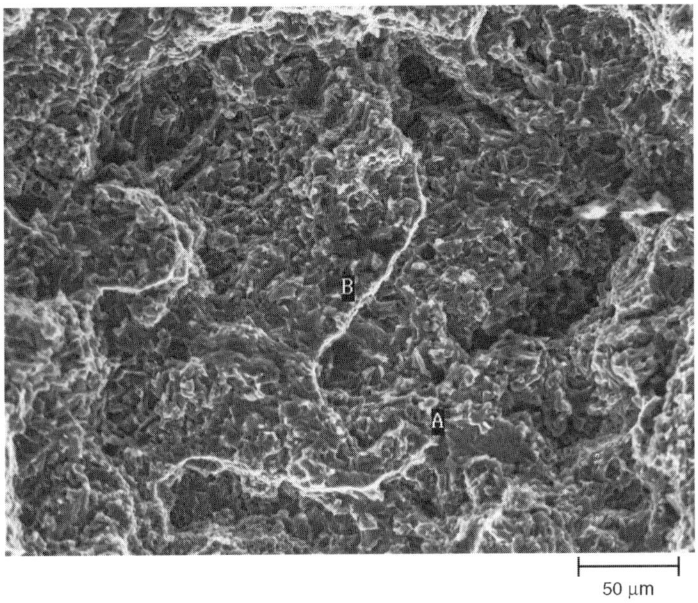

Fig. 4.6 Transcrystalline fracture of medium-developed surface. The tear ridges in the micronecks of α-aluminum solid solution have formed among the near eutectic grains. 300×

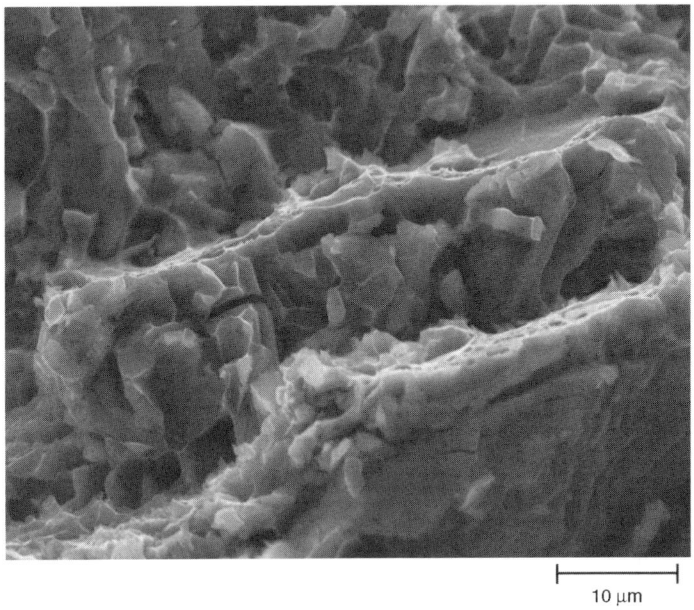

Fig. 4.7 Detail of [A] in Fig. 4.6. Two parallel micronecks have formed in the α-aluminum solid solution. The bands of the ductile dimples, visible in matrix are a result of its plastic microdeformation. 1800×

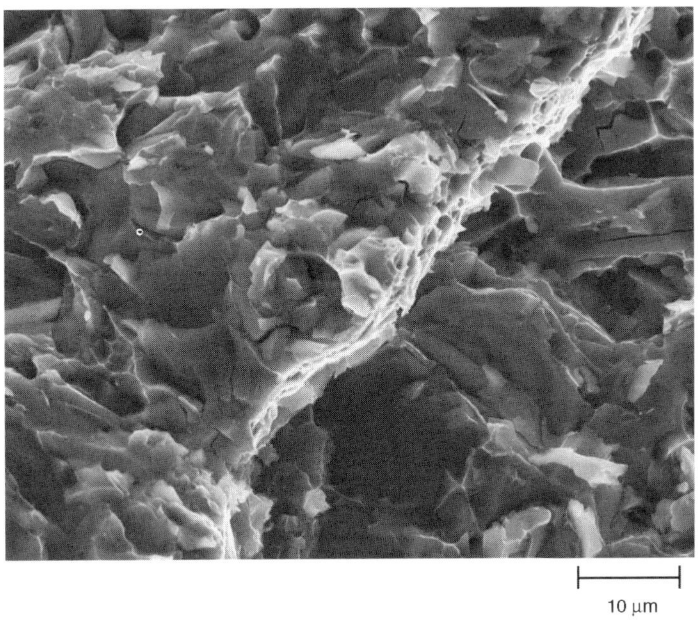

Fig. 4.8 Detail of [B] in Fig. 4.6. In the α-aluminum solid-solution plastic deformation took place resulting in the bands of the dimples formation. The cleavage crack crossed the silicon precipitates. 1500×

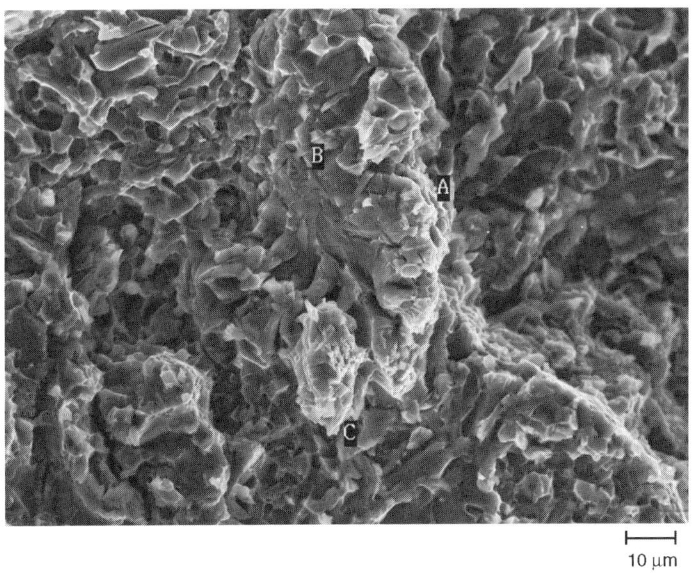

Fig. 4.9 Transcrystalline fracture of greatly developed surface. The crack crossed the eutectic zone α-Al + Si and the cleavage planes of the brittle eutectic phases. In the zones of the deformation in the matrix plastic fracture of the micronecks took place. 700×

Fig. 4.10 Detail of [A] in Fig. 4.9. The band of the open dimples is situated in the shear edge of the α-aluminum solid solution. 2600×

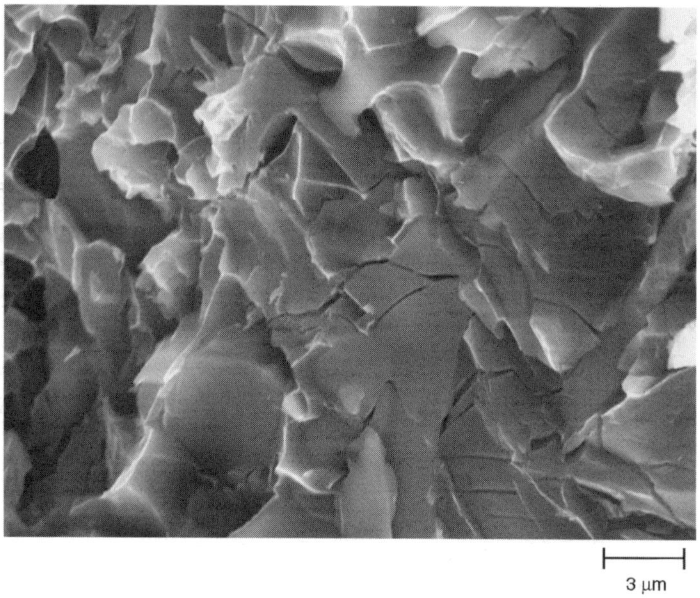

Fig. 4.11 Detail of [B] in Fig. 4.9. Transcrystalline, cleavage fracture. The crack crossed the cleavage planes of the brittle eutectic phases. Secondary cracks can be observed in this area. 2600×

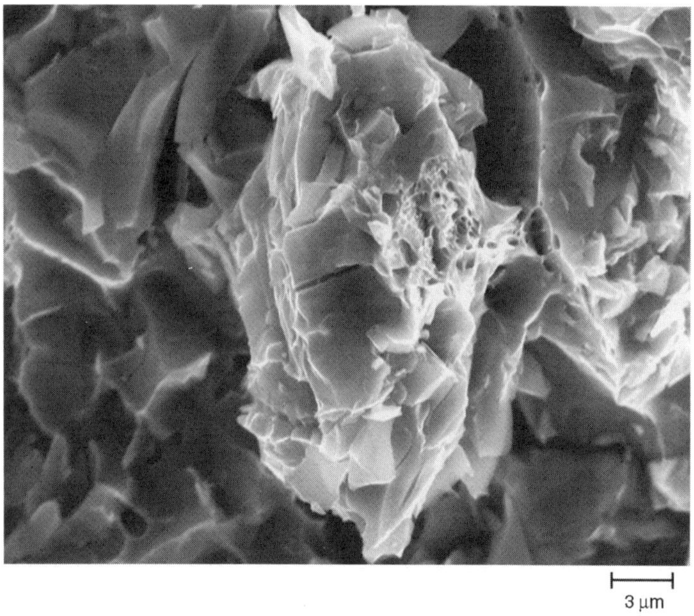

Fig. 4.12 Detail of [C] in Fig. 4.9. Transcrystalline fracture in the brittle eutectic phases with visible secondary cracks. In the center of the micrograph, the zone of the ductile, deformed matrix is shown. The dimple morphology is characteristic for a shear process. The dimples on the tear ridge in the micronecks of the α-aluminum solid solution have formed as a result of void coalescence. 3300×

Alloy 336.0 (AlSi13Mg1CuNi), Refined, Modified, Metal Mold Cast Part, Fracture after Static Tensile Test

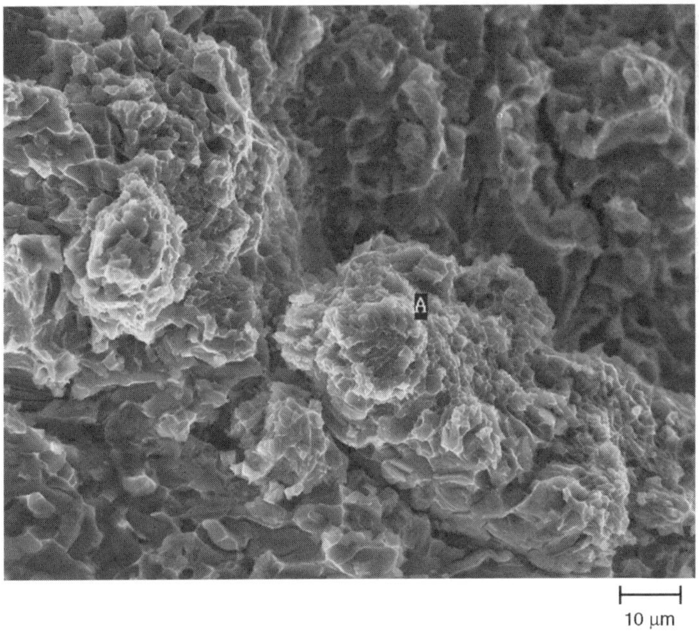

Fig. 4.13 Transcrystalline fracture of greatly developed surface. The main crack crossed the several eutectic grains (see Fig. 4.4). 900×

Fig. 4.14 Detail of [A] in Fig. 4.13. Transcrystalline fracture with visible cells of different morphology. Some cells were formed around the small eutectic silicon particles (center of the micrograph). 4500×

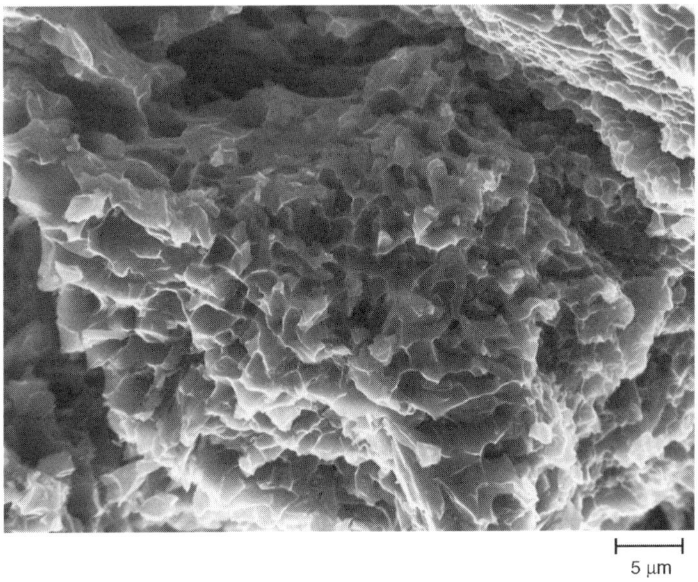

Fig. 4.15 Transcrystalline, cellular fracture in the eutectic α-Al + Si grain. 2000×

Fig. 4.16 Fracture of mixed morphology. Cleavage transcrystalline and cellular. The crack crossed the cleavage planes in the eutectic silicon precipitates. Among the cracked silicon particles the micronecks of the α-aluminum solid solution can be observed. The local tear ridges were formed as a result of the local plastic deformation in the α-aluminum solid solution. 2000×

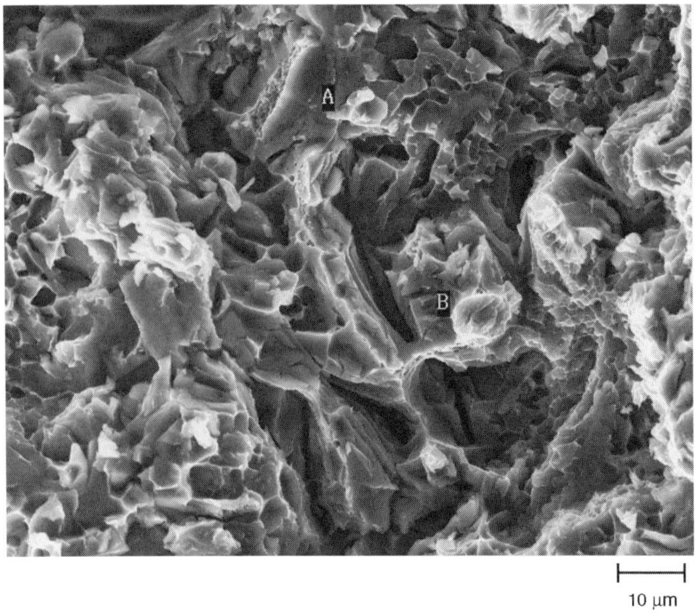

Fig. 4.17 Transcrystalline fracture. The deep secondary cracks are present in the brittle eutectic phases. The ductile fracture with the bands of the dimples can be observed in the matrix (bands of dimples). 950×

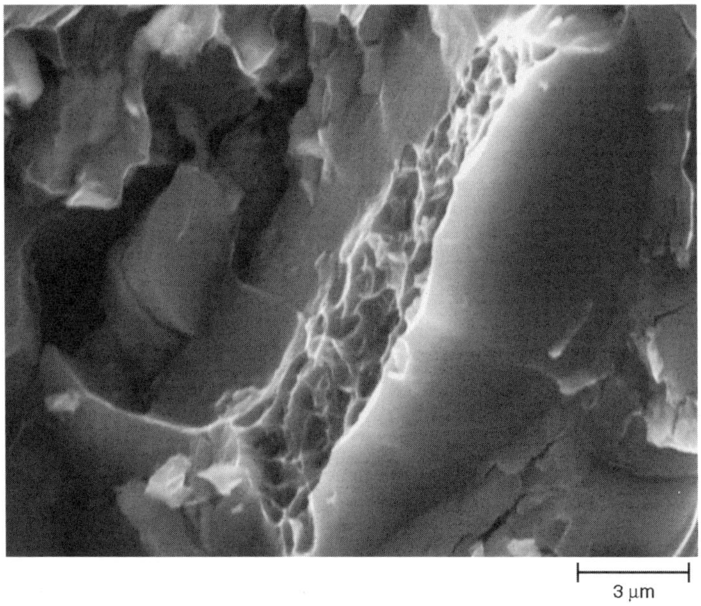

Fig. 4.18 Detail of [A] in Fig. 4.17. The micronecks in the deformed matrix area cracked along the tear ridge. On the interface between α-aluminum and silicon, cohesion was retained. The shear mechanism of decohesion in the matrix can be assumed. 5600×

Fig. 4.19 Detail of [B] in Fig. 4.17. Branched micronecks of the α-aluminum solid solution situated among the cracked Al_9FeNi intermetallic phase. The bands of the isolated dimples on the tear ridge are a result of formation and coalescence of the microvoids in the end stage of the fracture. 3000×

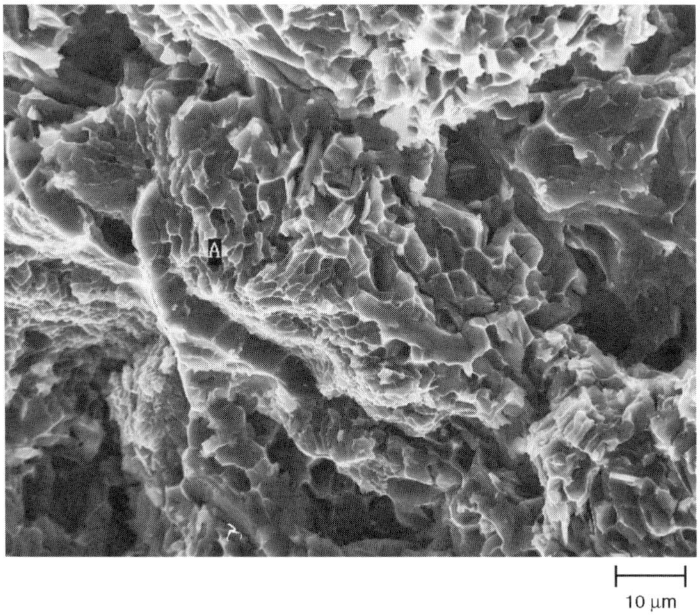

Fig. 4.20 Fracture in the boundary zone between eutectic grains. Fracture of cellular and ductile morphology is present in the two-phase regions. In the intermetallic Al_6NiCu_3 phase, the crack front crossed the cleavage planes. 1000×

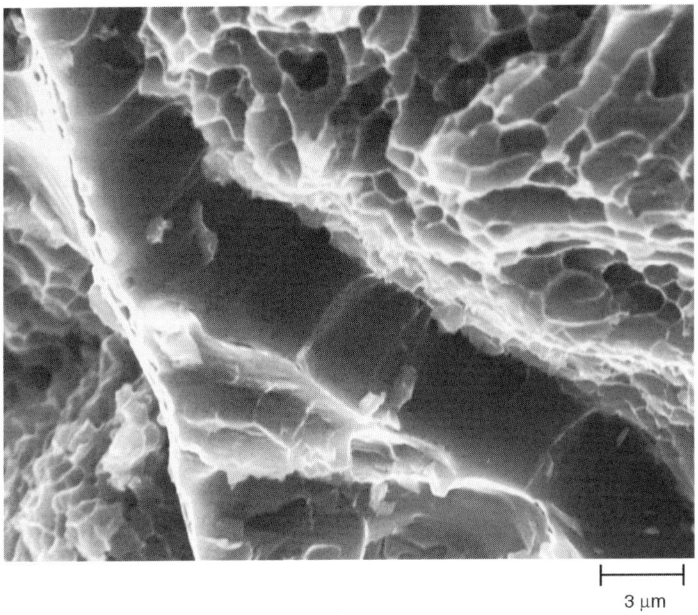

Fig. 4.21 Detail of [A] in Fig. 4.20. The edge of the microneck in the α-aluminum solid solution separates the eutectic zones. In the α-Al + Si eutectic, fracture of mixed morphology has formed. The part of the microneck situated in the center of the micrograph was plastically cracked. Oval, open dimples have formed after coalescence of the microvoids, by the shear fracture mechanism. The shear edges are shown. 4000×

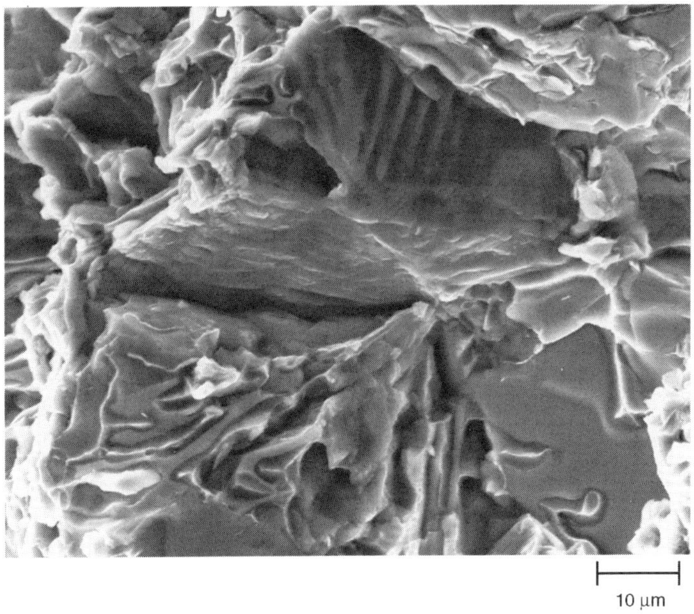

Fig. 4.22 Transcrystalline fracture in two-phase area α-Al+Al_6NiCu_3. In the precipitate of the brittle intermetallic phase, the crack front crossed the cleavage planes. In the matrix, the traces of plastic microdeformation (small dimples) can be observed. 1200×

CHAPTER 5

Alloy 355.0 (AlSi5Cu)

Composition and Properties

Alloy	Designation	Chemical composition (main components), wt%						Properties(a)	
		Si	Cu	Mg	Mn	Ni	Fe	R_m, MPa	A_5, %
As-examined	AlSi5Cu	4.5–5.5	1.0–1.5	0.35–0.6	0.2–0.5	. . .	0.9 max	200	0.5
Aluminum Association standard(b)	355.0	4.5–5.5	1.0–1.5	0.40–0.6	0.5 max	. . .	0.6 max	186	3

(a) R_m, ultimate tensile strength; A_5, elongation measured over a length of 5.65 S_o, where S_o is the cross-sectional area of the test specimen before the test. (b) Alloy 355.0 is registered with the Aluminum Association and designated by ASTM for sand and permanent mold casting. R_m for 355.0-T6 is the minimum ultimate tensile strength for a permanent mold casting for sample cut from casting (AMS 4281). T6 temper indicates the material has been solution heat treated by raising and holding the casting at a temperature long enough to allow the constituents to enter into solid solution. It is cooled rapidly so that the constituents remain in solution. Material is artificially aged to produce a stable temper but is not cold worked.

Microstructures

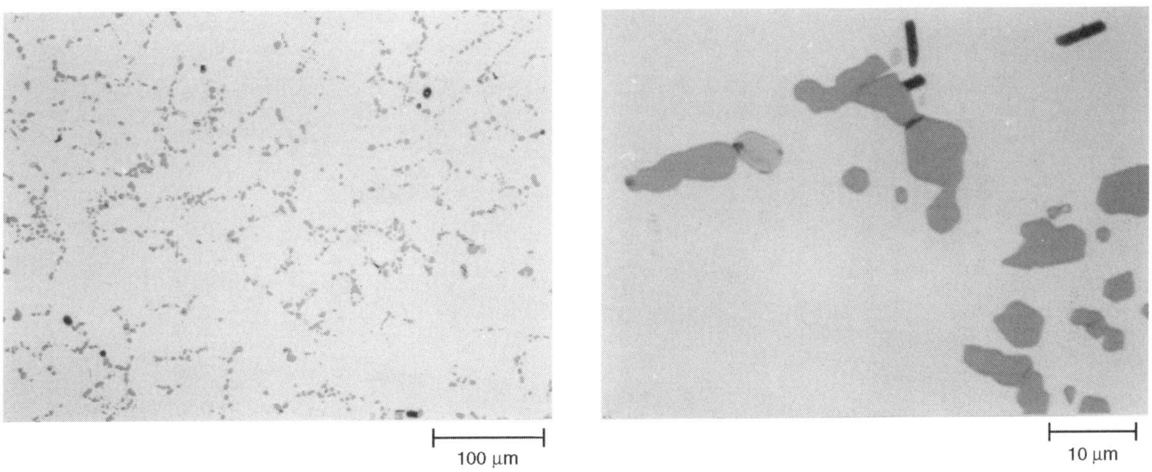

Fig. 5.1 Microstructures of AlSi5Cu (Alloy 355.0). Light microscope micrographs; etched with 1% HF. (a) After heat treatment (T6), 150×. (b) After heat treatment (T6), 1200×

Fracture Profiles of Alloy 355.0 (AlSi5Cu), Refined, Modified, T6, Permanent Mold Casting

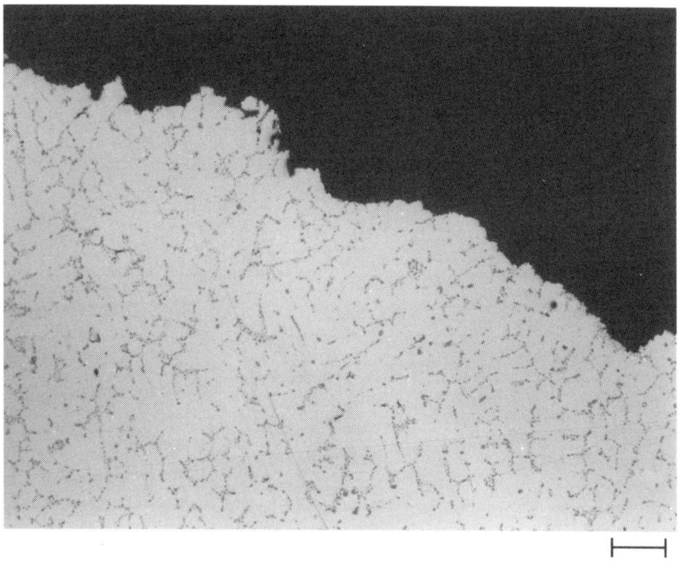

Fig. 5.2 Fracture profile of specimen after static tensile test. The main crack profile line reflects the morphology of the dendritic structure. 50×

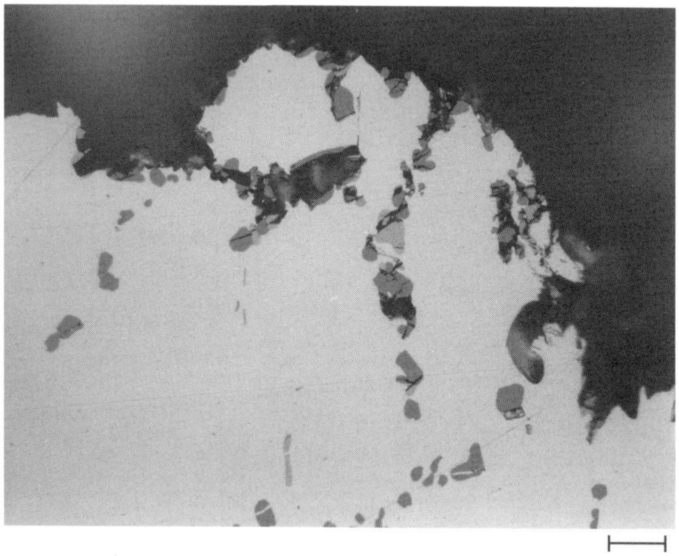

Fig. 5.3 Fracture profile of specimen after static tensile test. The main crack has formed in the polyphase region. Numerous cracks are present in the brittle particles of silicon and intermetallic phase. Among them, the micronecks of the deformed α-aluminum solid solution have fractured. 400×

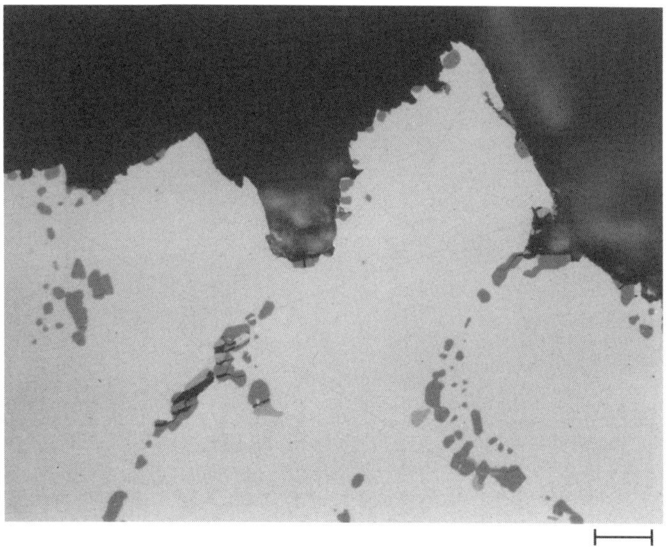

Fig. 5.4 Zigzag profile of the main crack in the specimen after static tensile test. The right part of each element is formed by the shear edge of the α-aluminum solid solution, while the left side is formed by the step profile. It was formed by the cleavage planes of the silicon particles and plastically deformed matrix zones (micronecks). In the zone near the fracture surface, secondary cracks in the brittle phases are visible. 400×

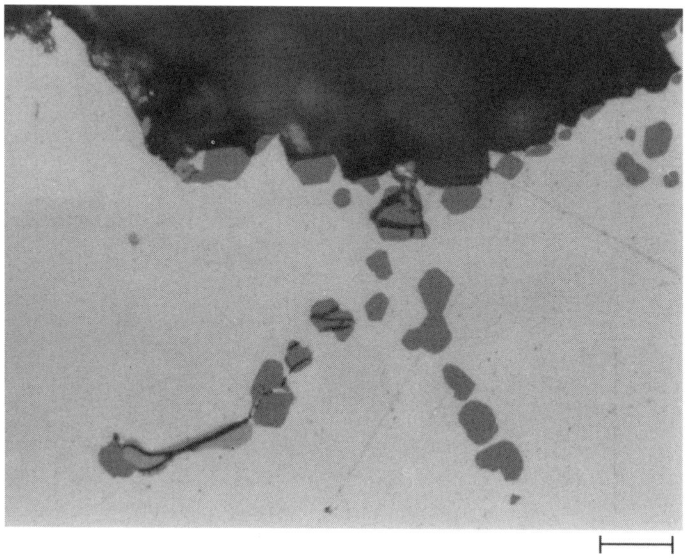

Fig. 5.5 Fracture profile of specimen after static tensile test. The main crack crosses a two-phase region. It has propagated on the parallel cleavage planes in the silicon particles and the plastically deformed microregions of the matrix. Secondary cracks are visible in the silicon and the brittle intermetallic phase precipitates. 1000×

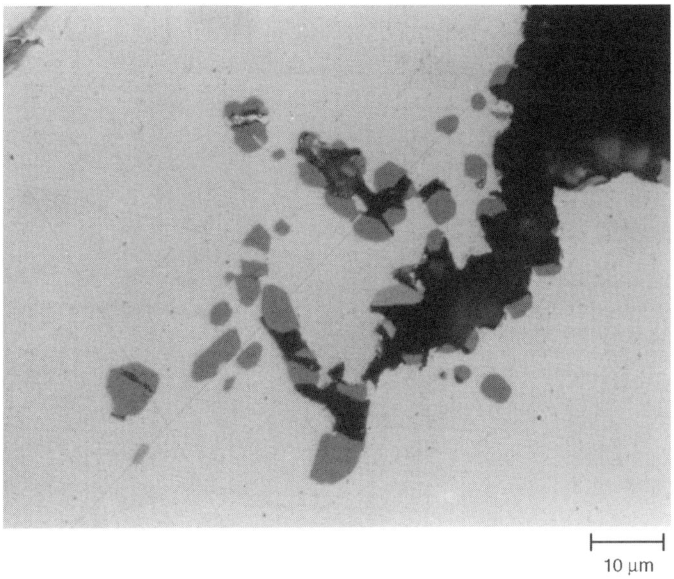

Fig. 5.6 Fracture profile of specimen after static tensile test. Brittle secondary cracks in the silicon and the intermetallic $Al_5Cu_2Mg_8Si_6$ phase precipitates are visible. 1000×

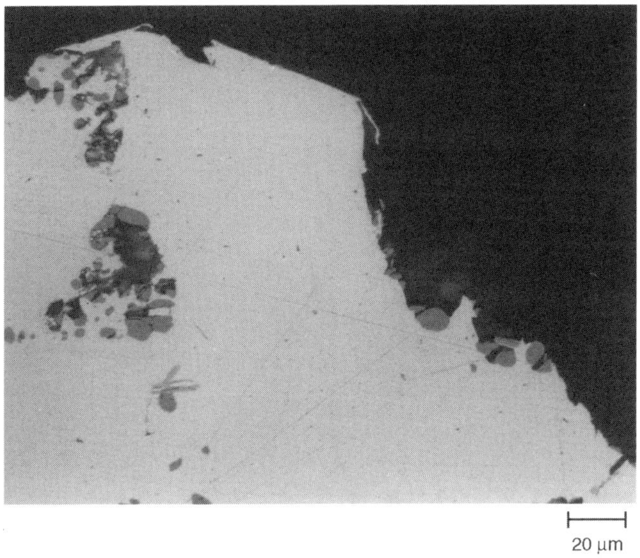

Fig. 5.7 Fracture profile of specimen after static tensile test. The profile of the main crack presents the shear edge of the α-aluminum solid solution. In the two-phase regions, the profile line is formed by the cleavage cracks in silicon particles and the short micronecks of the α-aluminum solid solution. In the brittle particles the secondary cracks are visible. 400×

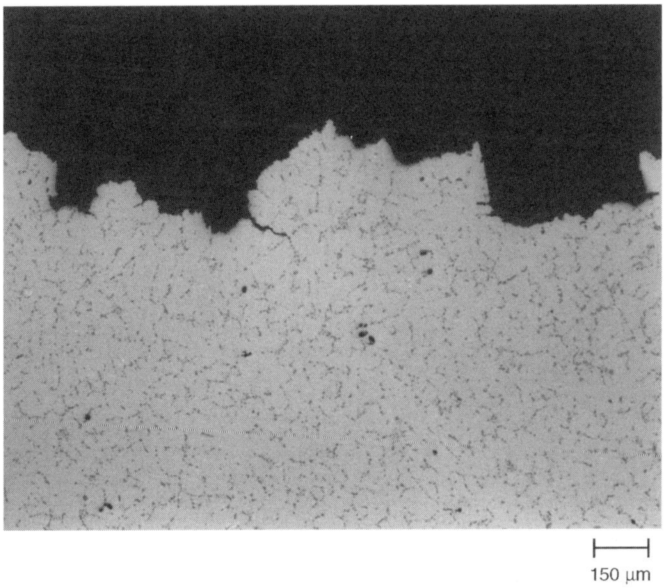

Fig. 5.8 Fracture profile of specimen after V-notch impact test at room temperature. The main crack is formed in the two-phase region. Secondary cracks and shear edges in several solid solution dendrites are visible. 50×

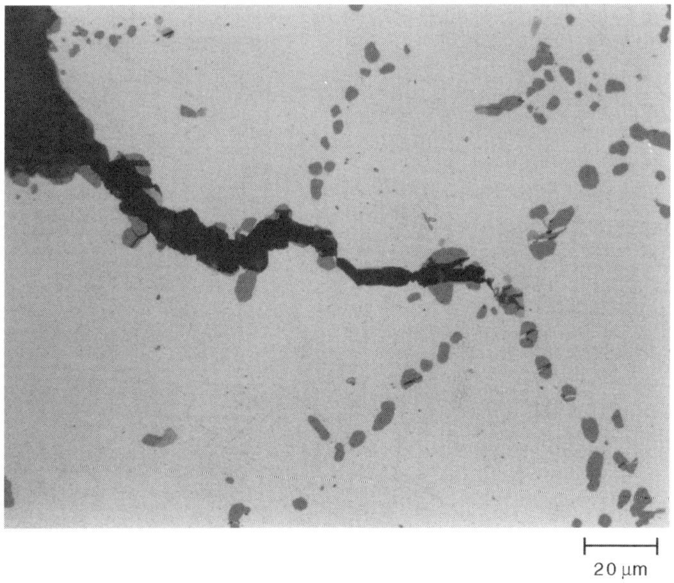

Fig. 5.9 Detail of the profile visible in Fig. 5.8. Secondary cracks in the polyphase region and in the silicon particles are visible. 500×

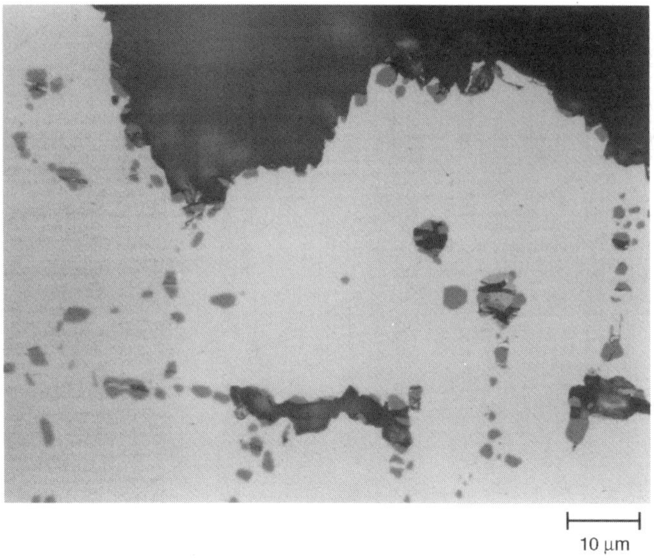

Fig. 5.10 Fracture profile of specimen after V-notch impact test at room temperature. The main crack crossed the two-phase region. Cleavage lines in silicon particles and the shear edges in the microregions of the matrix form the elements of the step on the profile line. Numerous brittle secondary cracks are visible. 1000×

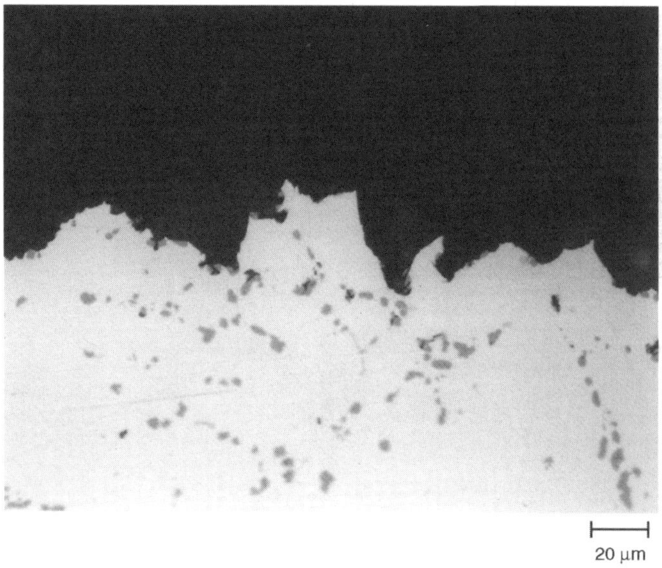

Fig. 5.11 Fracture profile of specimen after V-notch impact test at room temperature, central zone of the specimen. Profile of the main crack of zigzag shape is formed by the step elements. Shear edge lines and fractured micronecks of the α-aluminum solid solution are visible in two-phase regions as well. 400×

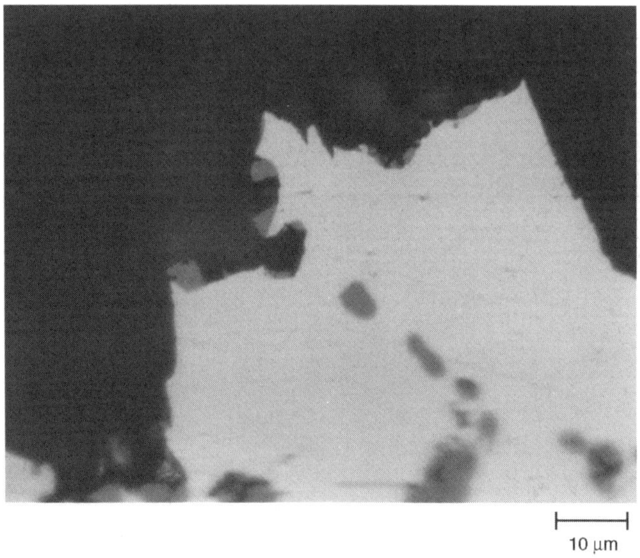

Fig. 5.12 Detail of the profile visible in Fig. 5.11. Two almost parallel lines of the shear of α-aluminum solid solution are ended with sharp fractured necks. The shear edge lines form the steps. The brittle cracks of silicon particles and the deformed microregions of matrix can be observed. The cohesion is retained on the α-Al/Si interface. 1000×

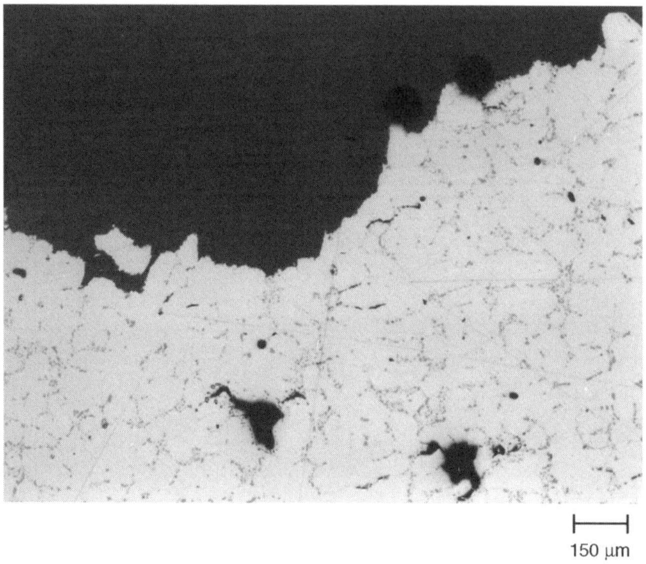

Fig. 5.13 Fracture profile of specimen after low-cycle fatigue test. The main crack crossed the two-phase region. Numerous brittle secondary cracks and shrinkage micropores are visible. 50×

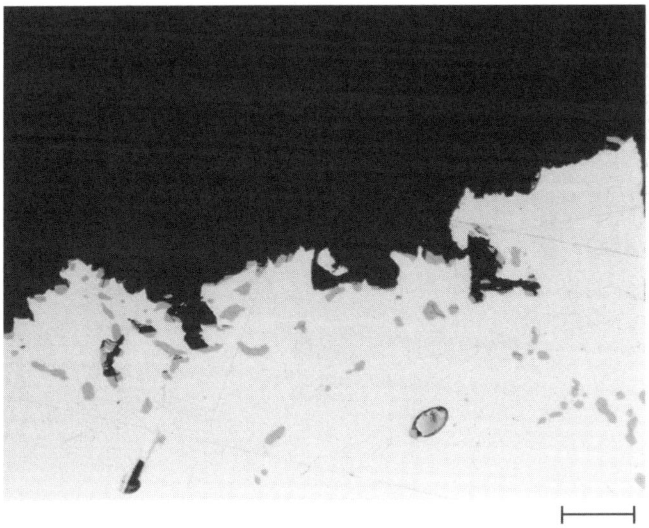

Fig. 5.14 Fracture profile of specimen after low-cycle fatigue test. The main crack crossed the polyphase region. Among cracked brittle silicon particles, the sharp micronecks of the deformed α-aluminum solid solution have cracked. 250×

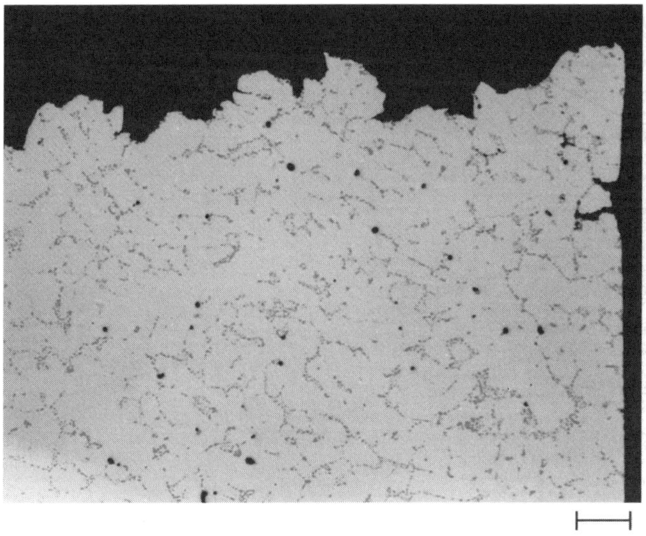

Fig. 5.15 Fracture profile of specimen after low-cycle fatigue test (rim zone of the specimen). The main crack is in the polyphase region. Secondary cracks in the polyphase regions are visible on the lateral surface of the specimen. 50×

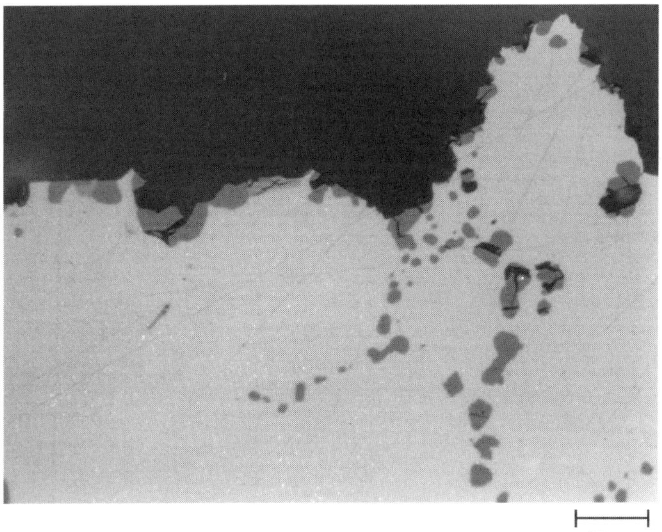

Fig. 5.16 Fracture profile of specimen after low-cycle fatigue test. The main crack crossed the polyphase region. The step line of the main profile is formed by the cleavage lines in the silicon precipitates and the shear edge lines in the matrix. 500×

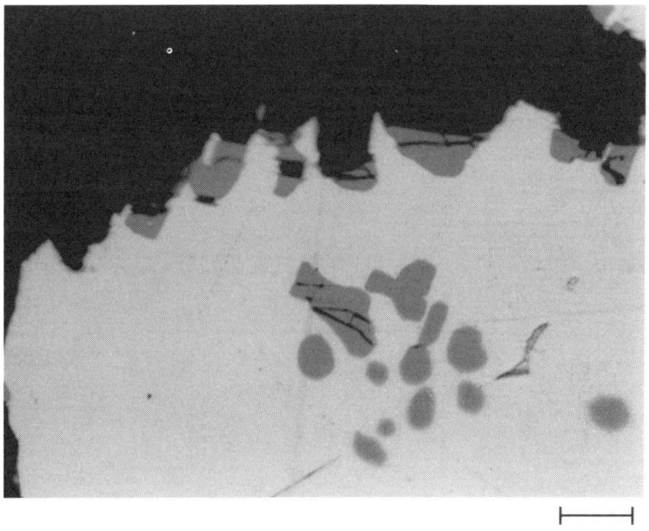

Fig. 5.17 Fracture profile of specimen after low-cycle fatigue test. The main crack is in the polyphase region. Cleavage lines in the silicon precipitates and the shear edge lines in matrix elements are forming the step line of the main profile. Numerous secondary cracks are present in the silicon particle. The decohesion zone on the interfaces between α-aluminum and silicon is visible. 1000×

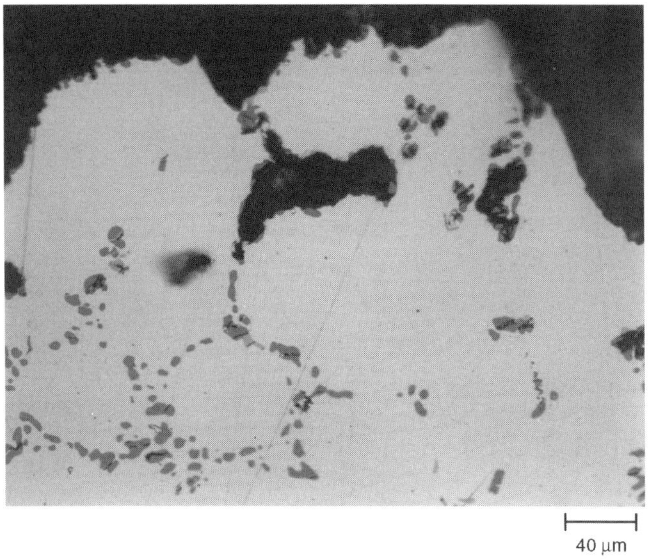

Fig. 5.18 Fracture profile of specimen after low-cycle fatigue test. In the polyphase region, the step profile is visible. The shear edge lines reflect the shear process in the α-aluminum solid solution. The shrinkage micropores are visible in the zone near to the fracture surface. 250×

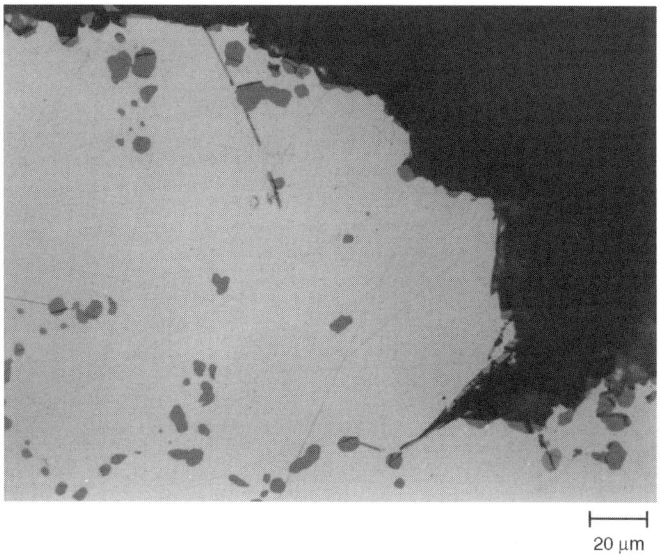

Fig. 5.19 Fracture profile of specimen after low-cycle fatigue test. The step line of the main profile reveals the cleavage cracks in the silicon and the intermetallic phase Al_7Cu_2Fe. 400×

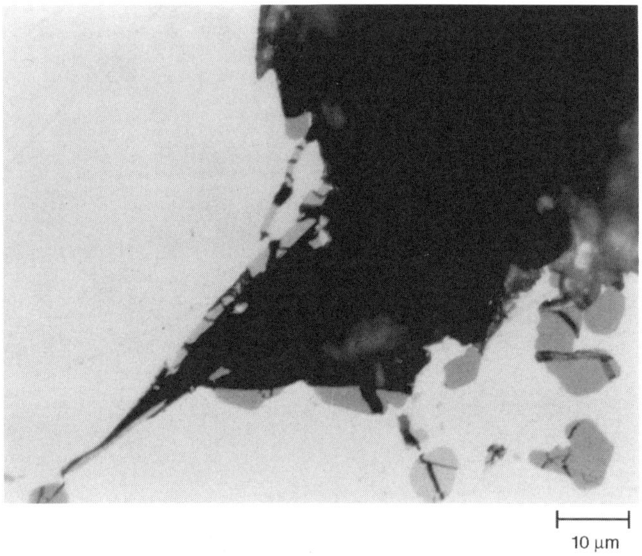

Fig. 5.20 Detail of the profile visible in Fig. 5.19. Numerous brittle cracks in intermetallic phase Al_7Cu_2Fe and in silicon are shown. 1000×

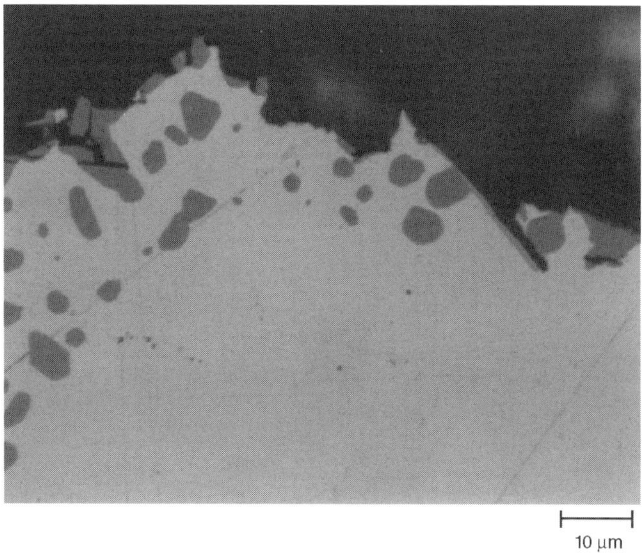

Fig. 5.21 Fracture profile of specimen after low-cycle fatigue test. In step profile line of the main fracture the secondary cracks in the needle precipitates of the Al_7Cu_2Fe and in the silicon particles are visible. Among the brittle particles the ligaments of the matrix are visible. 1000×

Fracture Surfaces of Alloy 355.0 (AlSi5Cu), Refined, Modified, Permanent Mold Casting, T6, Fracture after Static Tensile Test

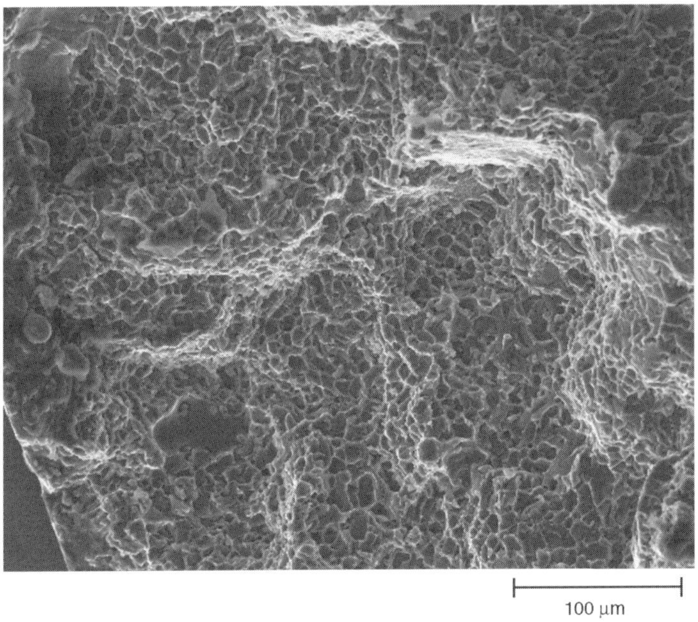

Fig. 5.22 Transcrystalline fracture in the rim zone of the specimen. The oxide inclusions are present on the fracture surface. The intercrystalline crack crossed the interface between α-aluminum and Al_2Cu. 250×

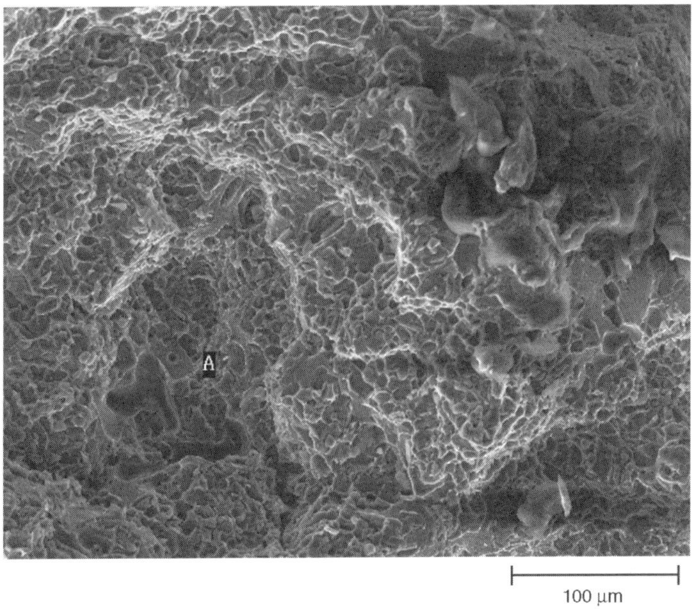

Fig. 5.23 Transcrystalline fracture in the specimen center zone. The oxide inclusions are visible on the fracture surface. The intercrystalline fracture was formed on the interface between α-aluminum and Al_2Cu. 250×

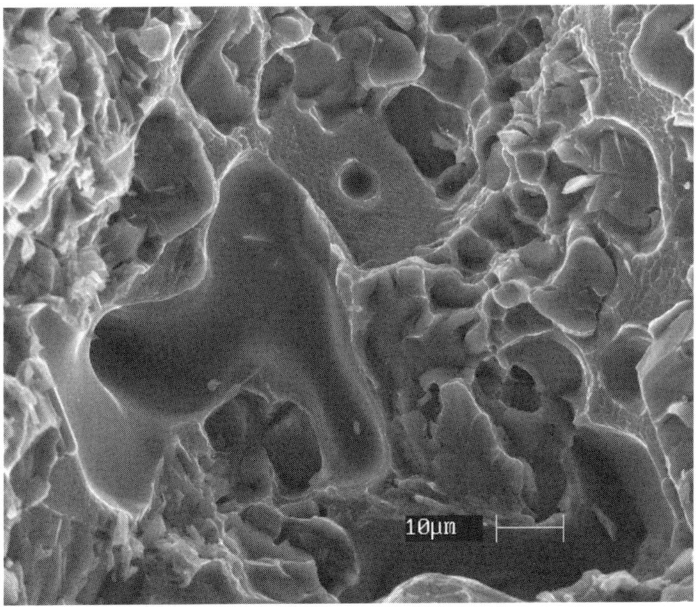

Fig. 5.24 Detail of [A] in Fig. 5.23. Fracture of mixed morphology: intercrystalline fracture on the interface between α-aluminum and Al_2Cu; transcrystalline, cleavage fracture in the $Al_5Cu_2Mg_8Si_6$ phase and in the silicon; and ductile fracture in the α-aluminum solid solution. The micronecks in the deformed α-aluminum solid solution regions are visible. 1000×

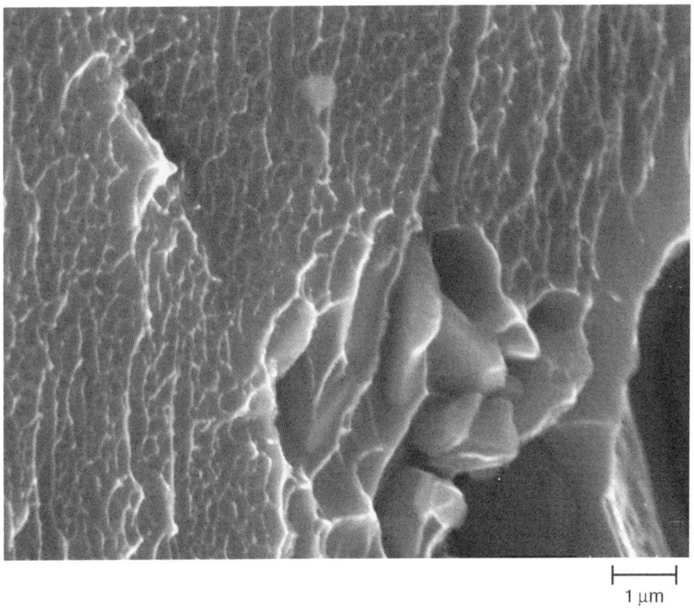

Fig. 5.25 Fracture morphology in the shear region of the of the α-aluminum solid solution. The oval dimples are visible. They point out into direction of the crack front propagation. In the silicon precipitate the cleavage crack on the several cleavage planes took place (see Fig. 5.5 and 5.6). 1000×

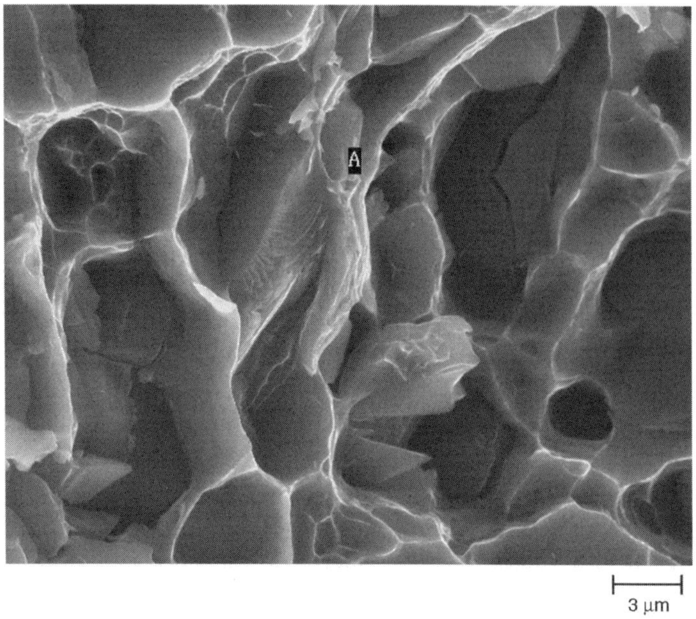

Fig. 5.26 Fracture of cellular morphology (see Fig. 5.6 and 5.7). Micronecks of the α-aluminum solid solution have formed around the silicon precipitates. In the silicon precipitates, secondary cracks are present. The interface cohesion was well retained. The dimples are a result of the local deformation in the α-aluminum solid solution. 3500×

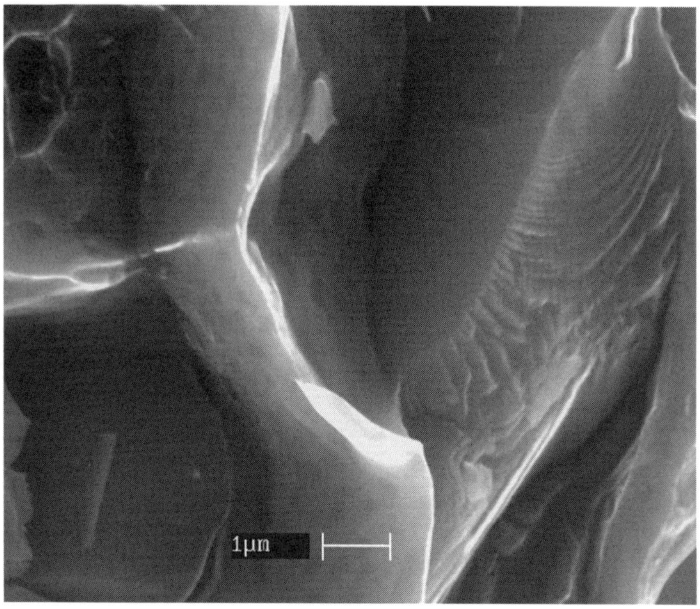

Fig. 5.27 Detail of [A] in Fig. 5.26. The tear ridge in α-aluminum solid solution was formed in the neighborhood of the cracked silicon precipitate. The step band has formed in silicon after the crack-propagation process was changed. The microregion visible in this micrograph is a single element of the step profile on the fracture profile line (see Fig. 5.6 and 5.7). 10,000×

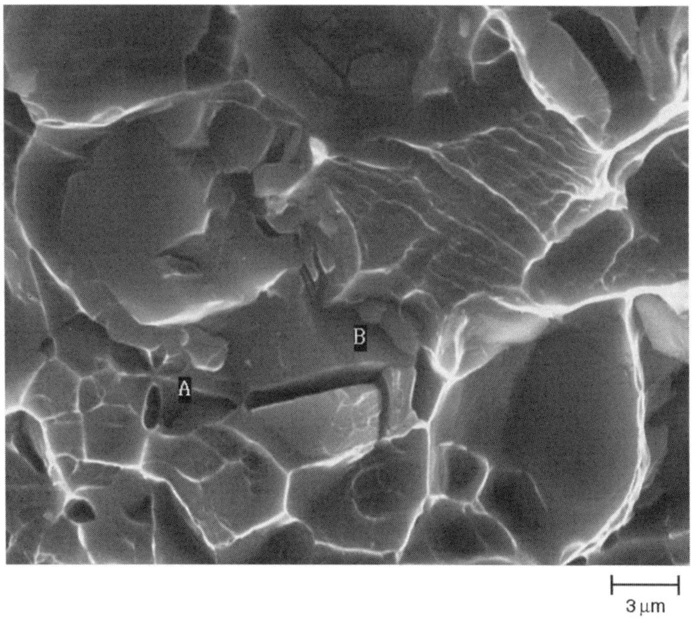

Fig. 5.28 Fracture morphology in the two-phase zone. In the silicon precipitates, the smooth cleavage facets are visible. The shear mechanism of fracture was active in the matrix. The oval dimples are present in this area, while on the microneck fracture edges the dimples are equiaxial. The zones of the retained interface cohesion can be observed. 3500×

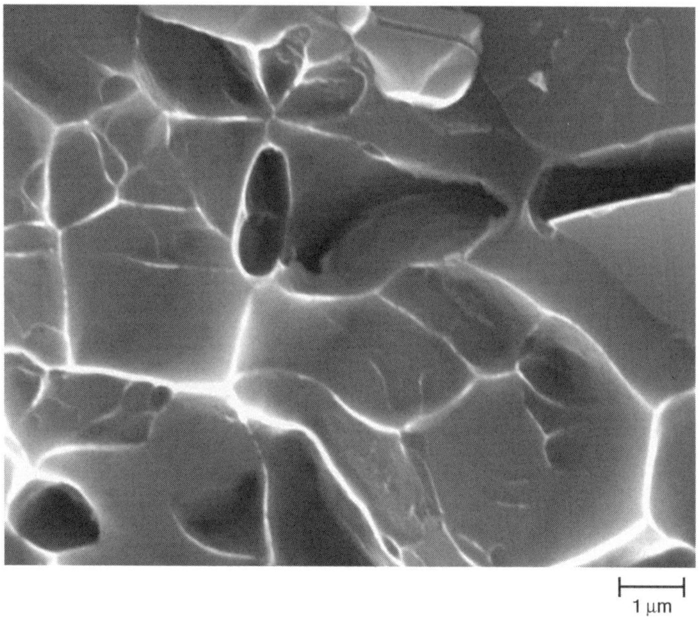

Fig. 5.29 Detail of [A] in Fig. 5.28. Secondary crack formed in the silicon precipitate. The tongue was formed on the cleavage facet after the displacement of the crack front by the zone of the microdeformation of the crystal lattice (top right of micrograph). The oval and open dimples and small tear ridges can be observed in the matrix. 10,000×

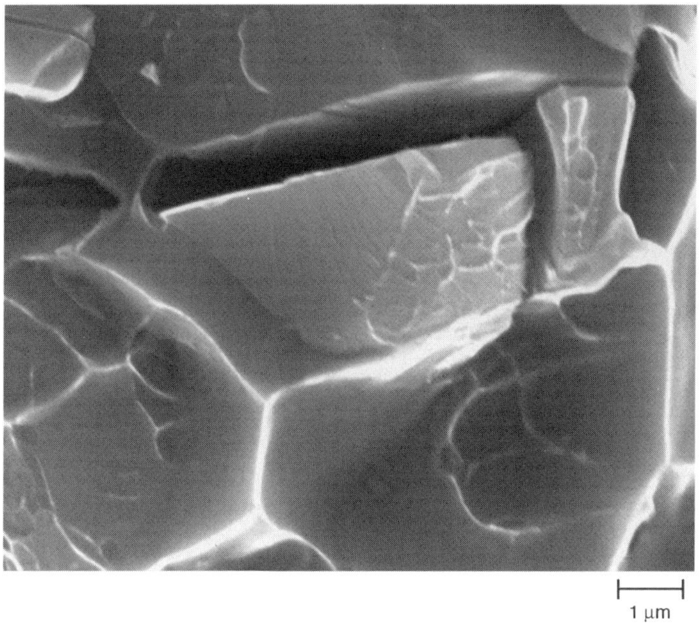

Fig. 5.30 Detail of [B] in Fig. 5.28. Rectilinear, secondary crack in the silicon precipitate. Tongues and cleavage steps are visible on the cleavage facet. The interface cohesion was retained on the interface between α-aluminum and silicon. 10,000×

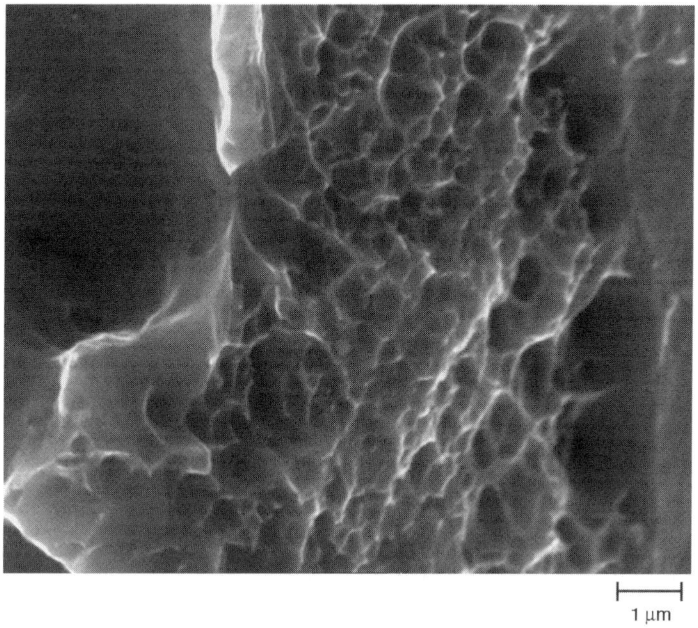

Fig. 5.31 Morphology of the fracture in the shear zone in the matrix (see Fig. 5.4). Dispersed particles of the intermetallic phase are visible in the small, ductile dimples. In these points the initiation of void formation took place. 10,000×

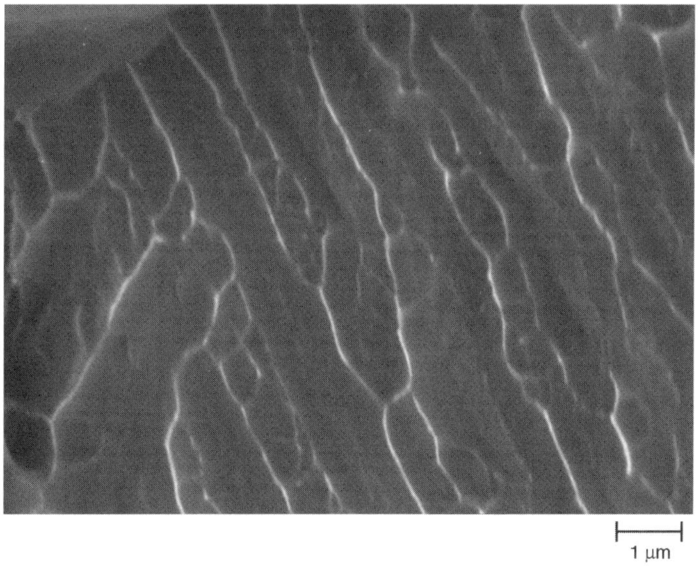

Fig. 5.32 Shear dimples arranged into bands on the fracture surface in the matrix (see Fig. 5.7). 10,000×

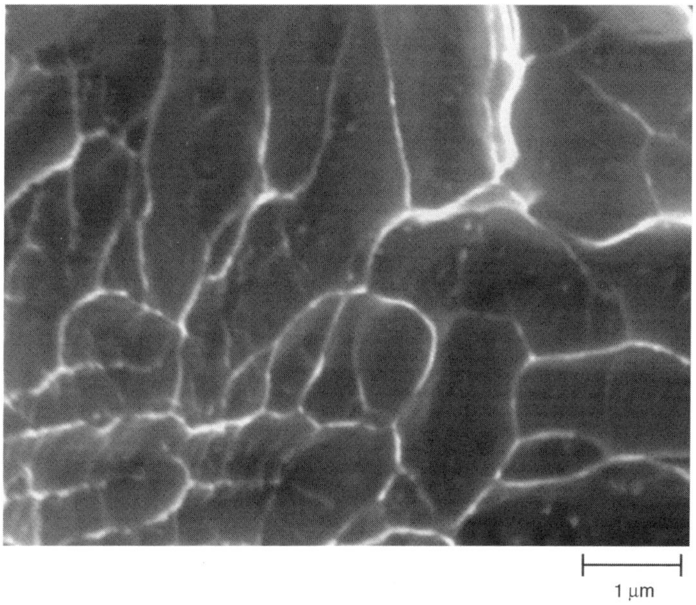

Fig. 5.33 Dimples in the matrix as a result of the void coalescence process around dispersed particles of the intermetallic phases. 15,000×

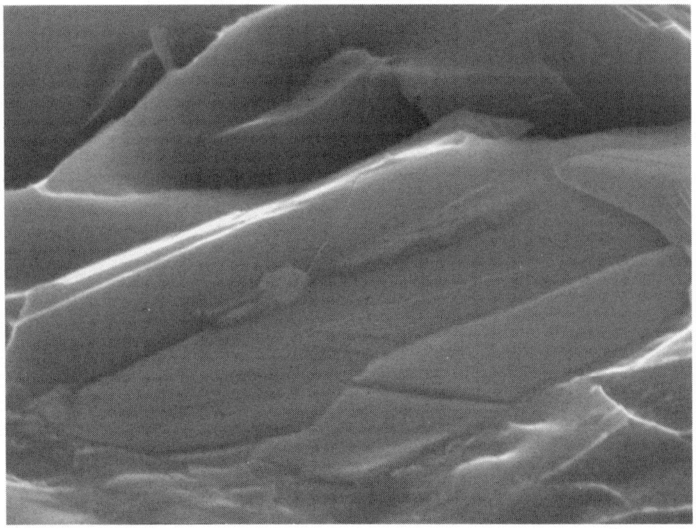

Fig. 5.34 Fracture surface in the silicon precipitate. The crack front crossed several cleavage planes, separated with the cleavage steps. 10,000×

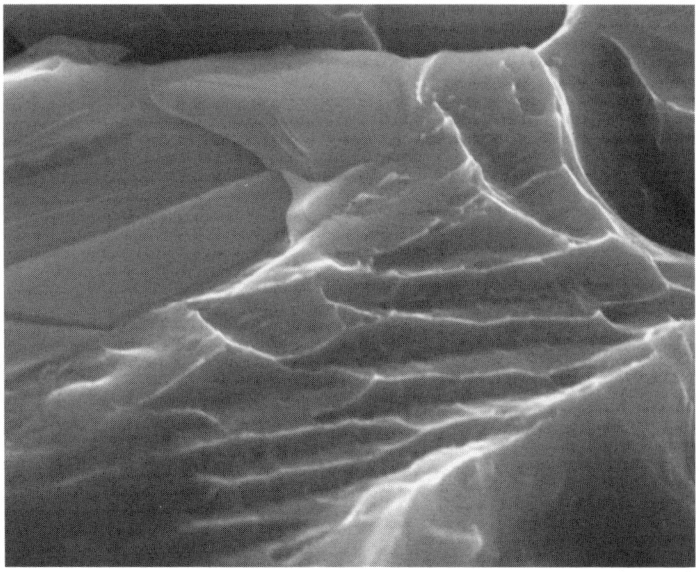

Fig. 5.35 Fracture surface in the α-aluminum solid solution between two silicon particles. The parallel tear ridges are visible. They are separated with the void bands, formed around dispersed particles of the intermetallic phase (the part of the fracture profile visible in Fig. 5.6). 10,000×

Fracture Surfaces of Alloy 355.0 (AlSi5Cu), Refined, Modified, Permanent Mold Casting, T6, Fracture after V-Notch Impact Test, at 21 °C (70 °F)

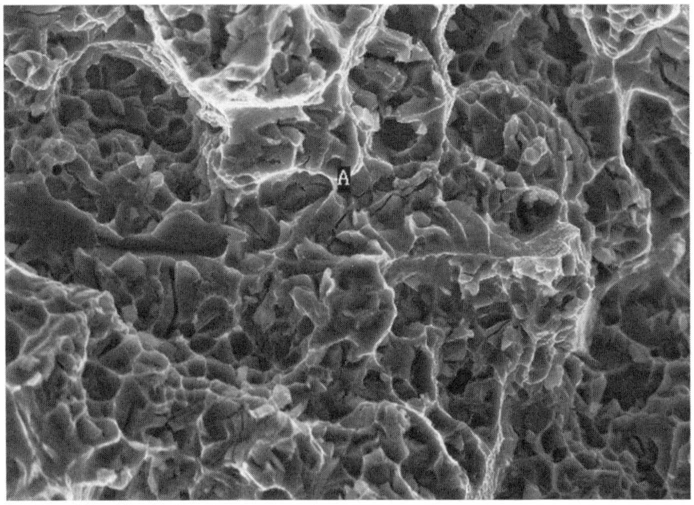

Fig. 5.36 Fracture morphology in the polyphase region of a greatly developed surface. Cleavage cracks are visible in the brittle eutectic phases. The micronecks were formed in the deformed α-aluminum solid solution (see Fig. 5.9). 500×

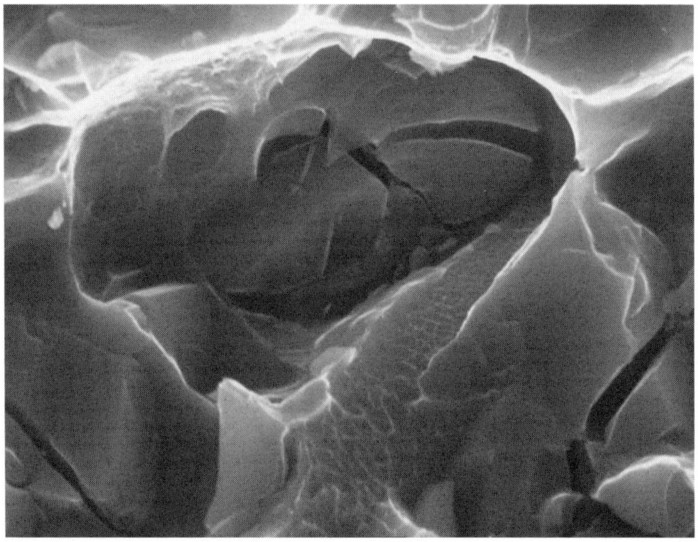

Fig. 5.37 Detail of [A] in Fig. 5.36. The cracks in the silicon particle are present. The steps have formed on the cleavage facets. The bands of the shallow dimples have formed on the shear surface in the matrix. 3500×

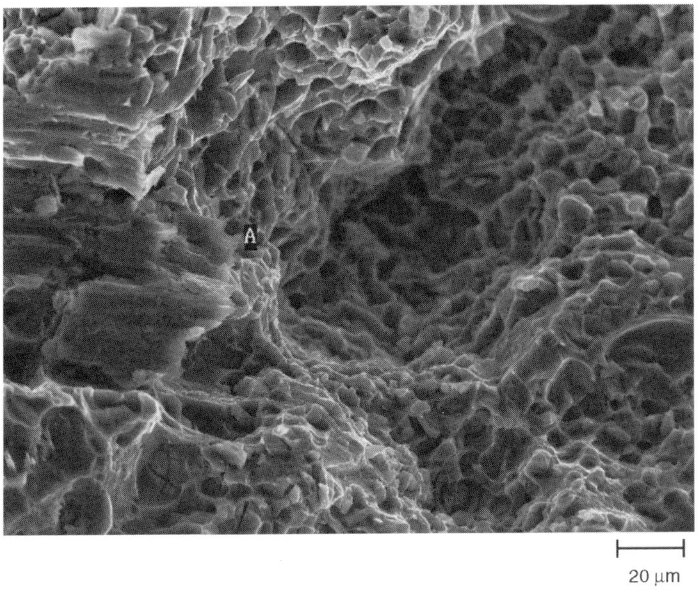

Fig. 5.38 Fracture surface morphology in the boundary zone between two areas of fracture—the shear matrix and the mixed two-phase region. 500×

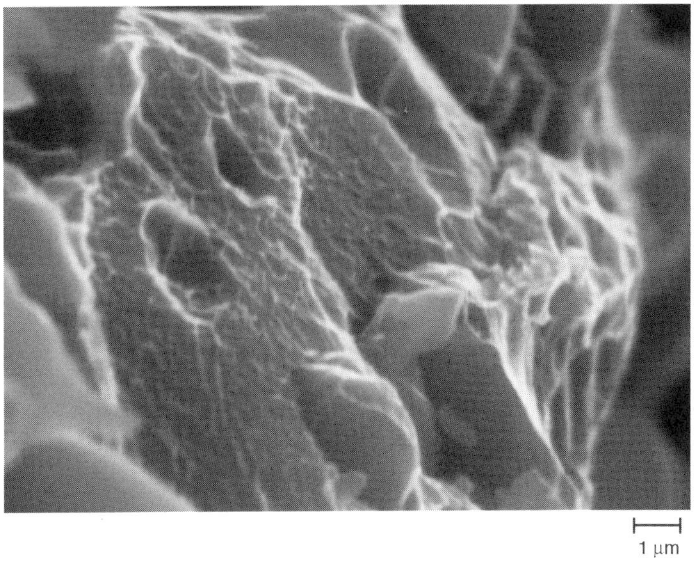

Fig. 5.39 Detail of [A] in Fig. 5.38. The shear surface in the α-aluminum solid solution, with the oval open, shear dimples, is situated between two silicon precipitates (see Fig. 5.12). 7000×

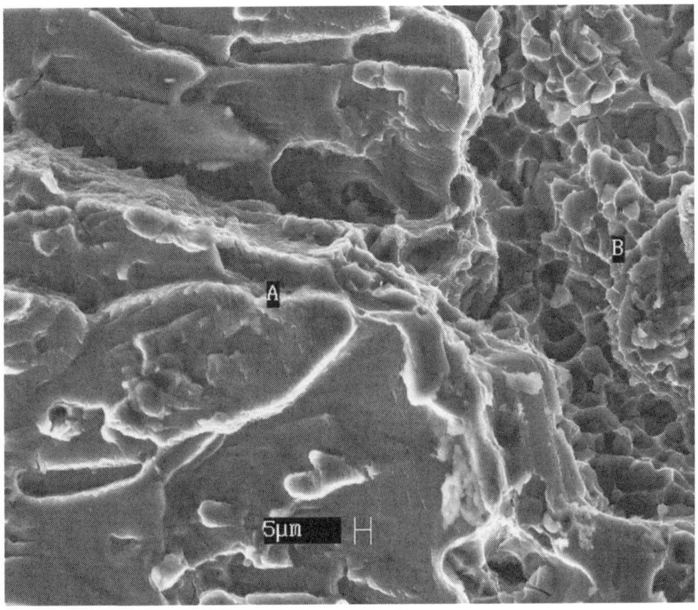

Fig. 5.40 The boundary zone: the shear fracture in the matrix and the cellular fracture in two-phase region (see Fig. 5.11). 500×

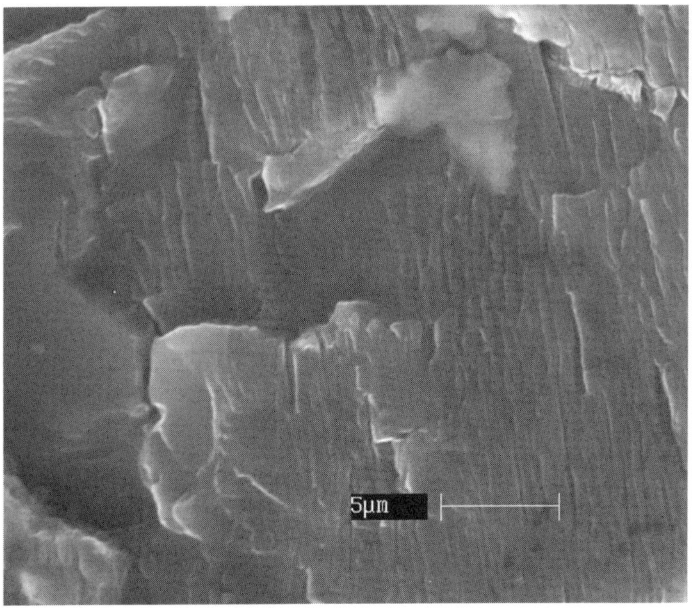

Fig. 5.41 Detail of [A] in Fig. 5.40. Morphology of the shear surface in matrix (see Fig. 5.12). 3600×

Fig. 5.42 Detail of [B] in Fig. 5.40. Morphology of the cellular fracture in two-phase region. Cleavage cracks are present in the brittle precipitates. They are round with the shear edge of the micronecks of the α-aluminum solid solution. 1500×

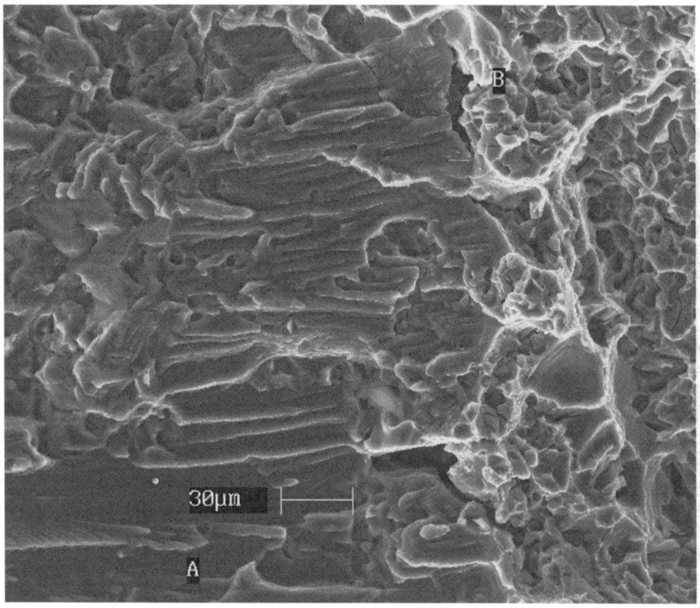

Fig. 5.43 The boundary zone between shear fracture in the matrix and cellular fracture in two-phase region. Secondary cracks can be observed. 350×

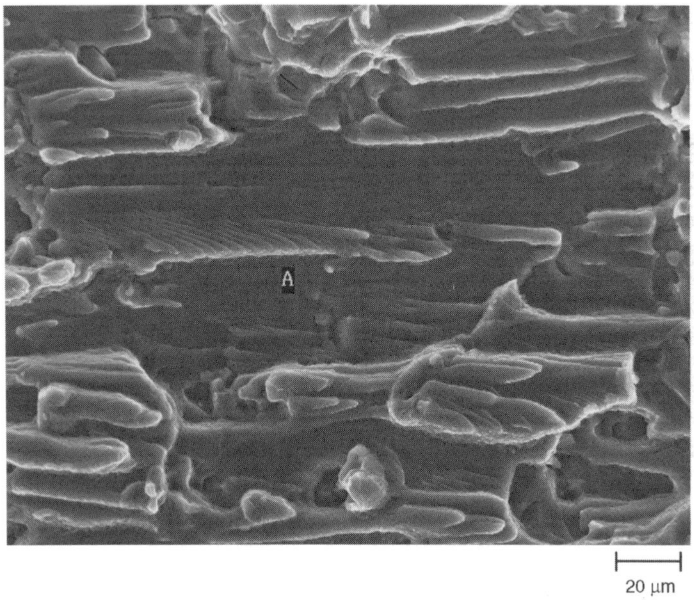

Fig. 5.44 Detail of [A] in Fig. 5.43. The shear bands are perpendicular to the zone boundary. 500×

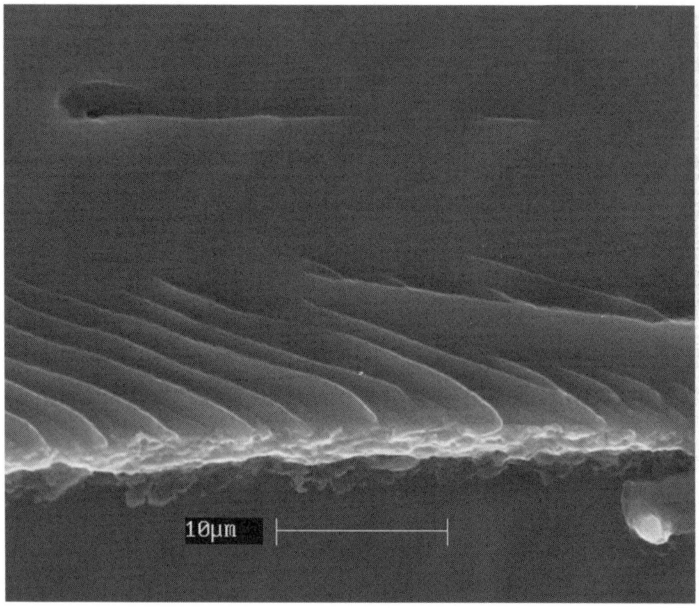

Fig. 5.45 Detail of [A] in Fig. 5.44. Morphology of the shear band. 2500×

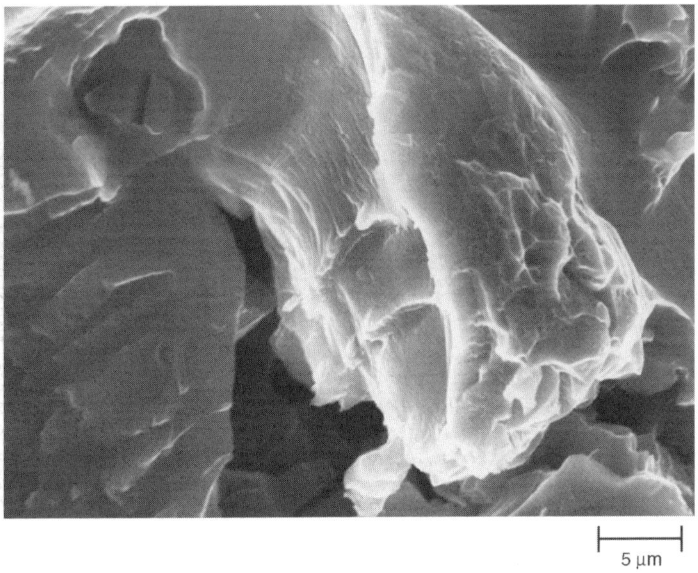

Fig. 5.46 Detail of [B] in Fig. 5.43. Secondary cracks in the boundary zone. The bands of the waved steps are parallel to the crack edge. In the microregion of the α-aluminum solid solution, traces of plastic deformation can be observed. 2600×

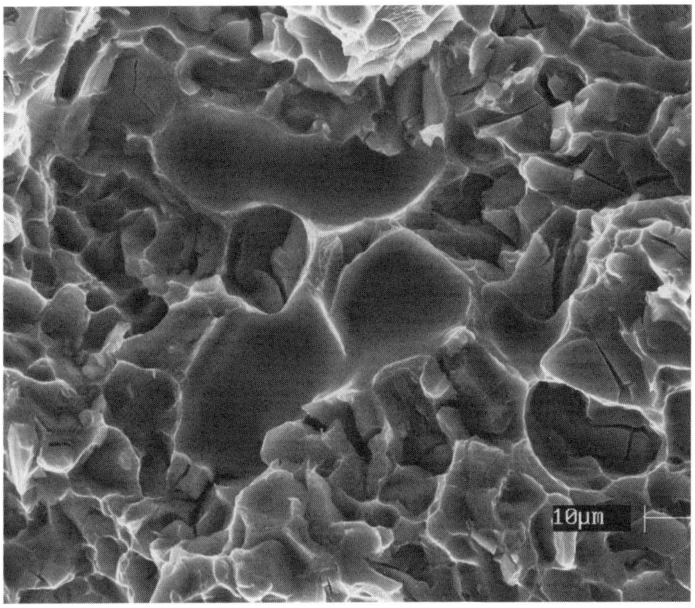

Fig. 5.47 Cellular fracture morphology in the polyphase region. The edges of the micronecks of the α-aluminum solid solution are present around fractured brittle eutectic phases. The intercrystalline fracture was formed on the interface between α-aluminum and $Al_5Cu_2Mg_8Si_6$. 1000×

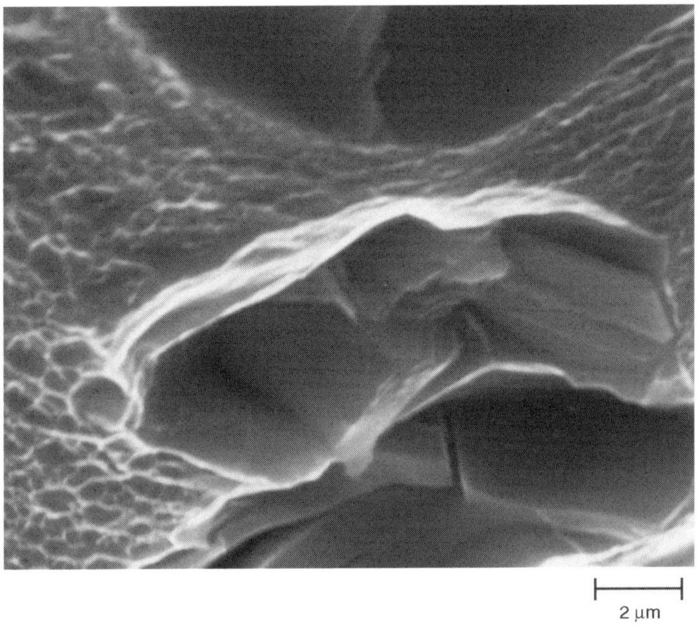

Fig. 5.48 The bands of dimples on the shear surface in the α-aluminum solid solution are situated between the cleavage cracks in the silicon particles (see Fig. 5.12). 6500×

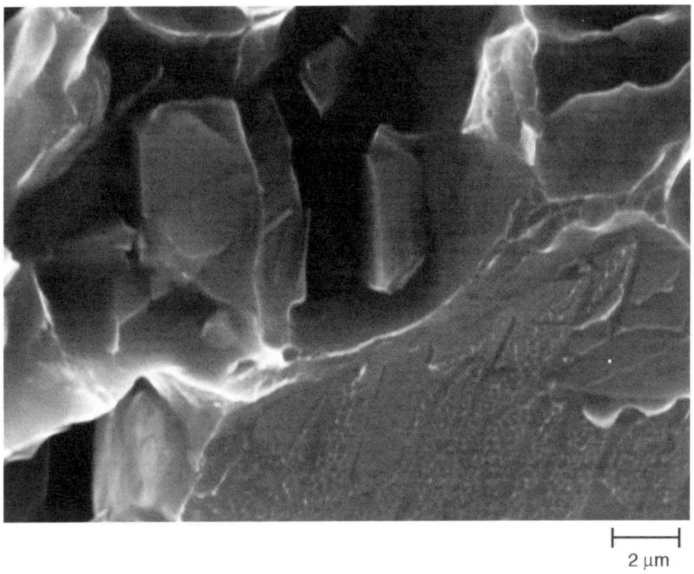

Fig. 5.49 Characteristic band distribution of dispersed particles of the intermetallic phase in the microdeformation zone of the α-aluminum solid solution (see Fig. 5.12) near the cracked silicon particle. The decohesion zone is situated on the interface between α-aluminum and silicon. 5000×

Fracture Surfaces of Alloy 355.0 (AlSi5Cu), Refined, Modified, Permanent Mold Casting, T6, Fracture after Low-Cycle Fatigue Test

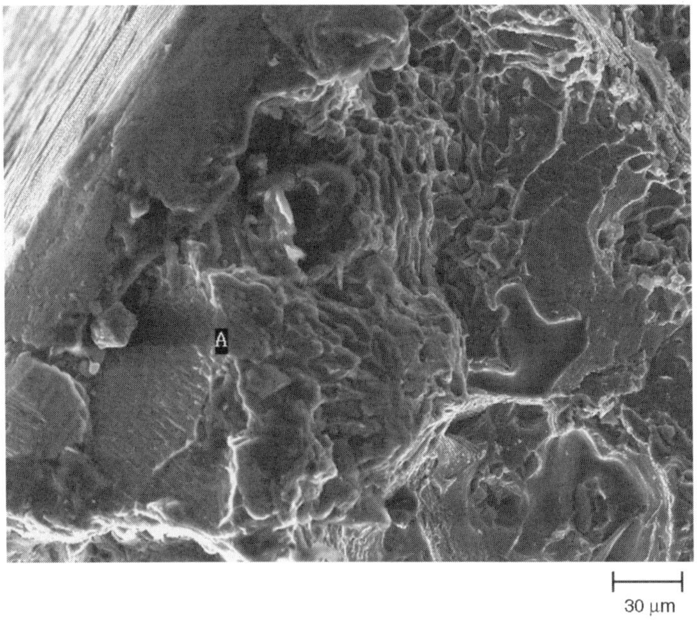

Fig. 5.50 Fracture in the rim of the specimen (see Fig. 5.15 and 5.16). Transcrystalline fracture surfaces of different morphology (cellular, cleavage, fatigue) are visible near the area of the intercrystalline fracture. The fraction of the deformed material (ductile fracture) increases when the distance from the specimen axis decreases. 350×

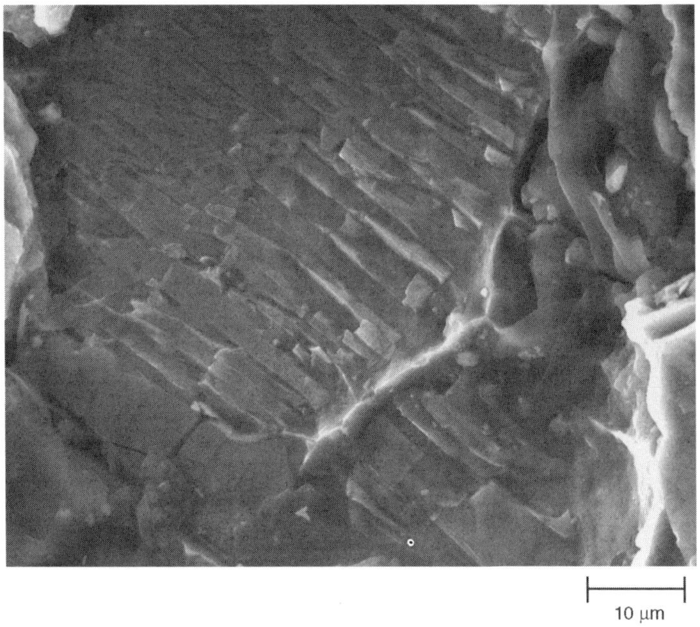

Fig. 5.51 Detail of [A] in Fig. 5.50. Fracture morphology in the rim zone of the specimen. In the matrix, the brittle fatigue striations are visible. 1500×

Fig. 5.52 Radial system of the cracks, propagated in the silicon crystal. 4000×

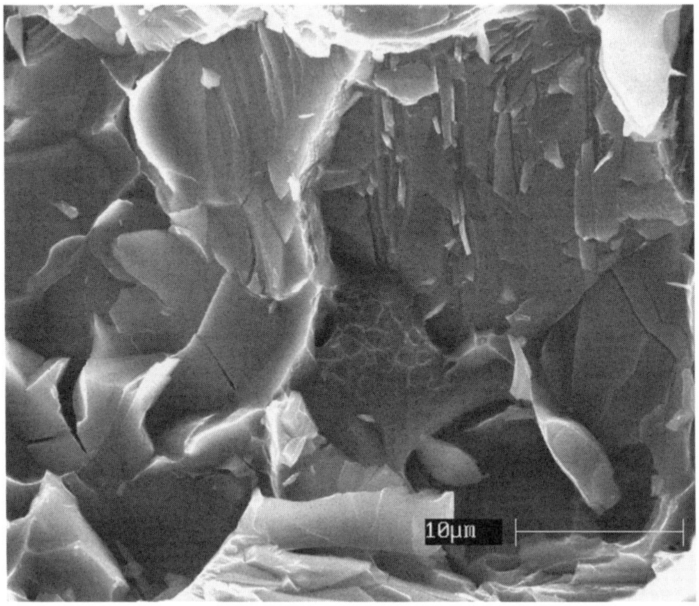

Fig. 5.53 Cellular fracture morphology in two-phase zone. The crack crossed several silicon precipitates (see Fig. 5.18). Micronecks in the α-aluminum solid solution are around the silicon particles. In one of them, the band of steps was formed as a result of the cleavage crack on several cleavage planes. 2000×

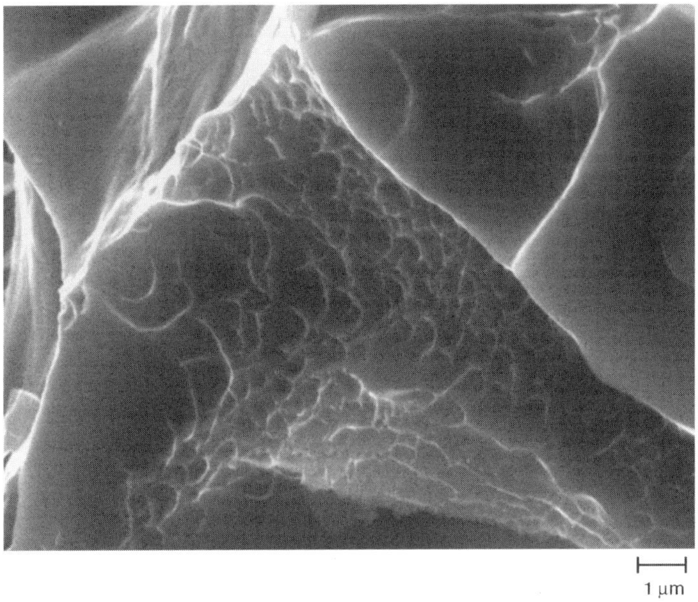

Fig. 5.54 Fracture morphology in the shear zone of the matrix (see Fig. 5.18). The region of plastic microdeformation can be observed. The shear dimples point out the crack front propagation direction. 7000×

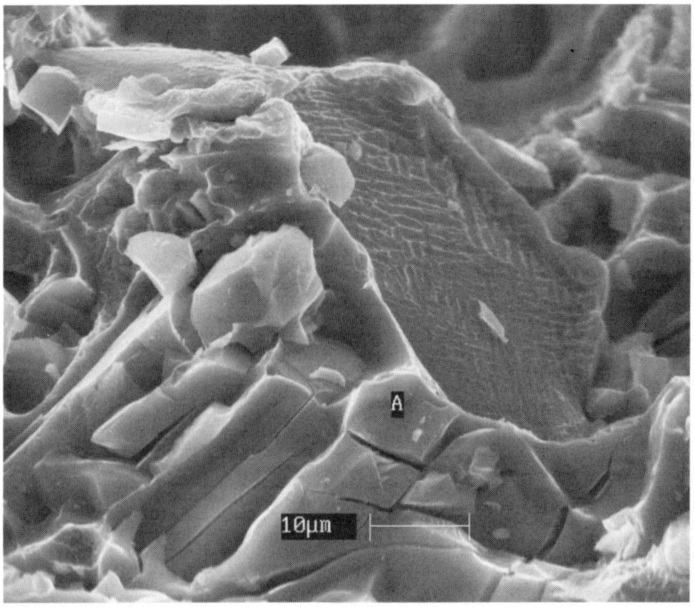

Fig. 5.55 Fracture of cellular morphology in the polyphase region. In the matrix, the bands of the dimples caused by plastic deformation are present. Secondary cracks have formed in the silicon precipitates. 1500×

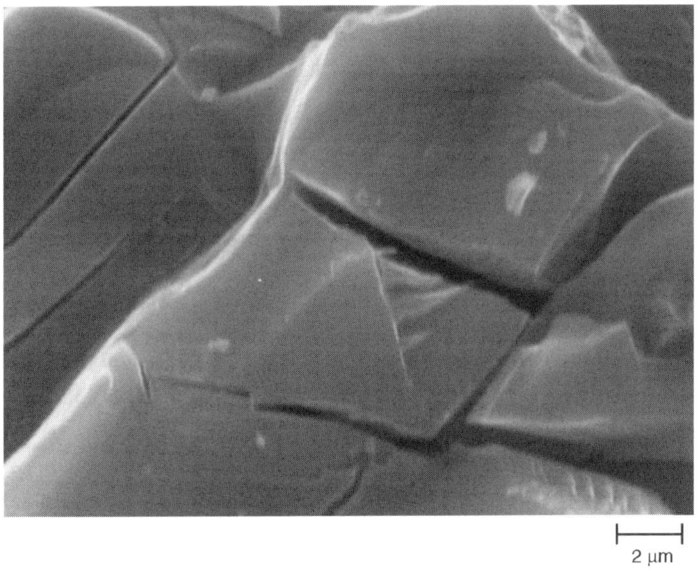

Fig. 5.56 Detail of [A] in Fig. 5.55. Secondary cracks and the branched steps in silicon crystal (see Fig. 5.17 and 5.18). 5000×

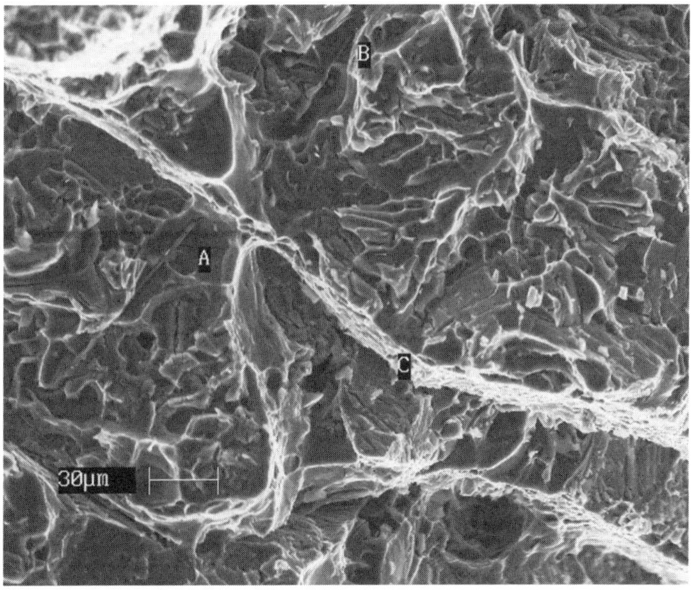

Fig. 5.57 Transcrystalline fracture with weakly developed surface. The regions of the composed morphology are separated by the edge of the shear zone in the α-aluminum solid solution. Al_7Cu_2Fe phase precipitates in the shape of needles can be observed at [A] and [B]. 450×

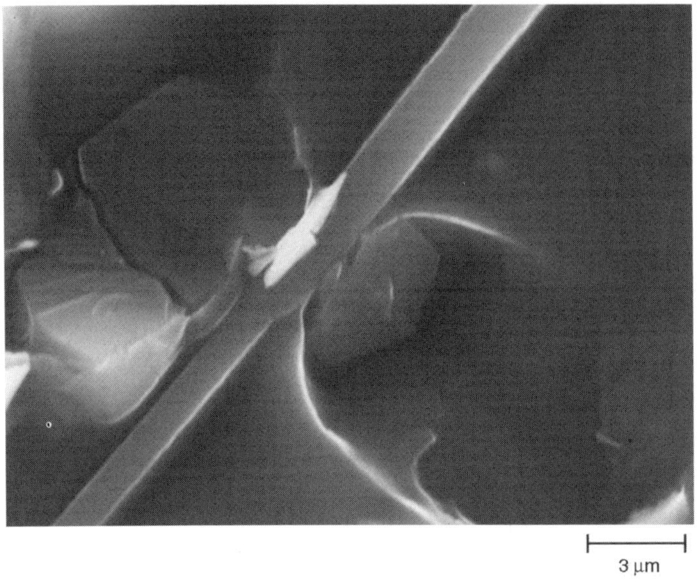

Fig. 5.58 Detail of [A] in Fig. 5.57. The needle-shape precipitate of the Al_7Cu_2Fe phase is situated between two oval dimples, around silicon precipitates. 5000×

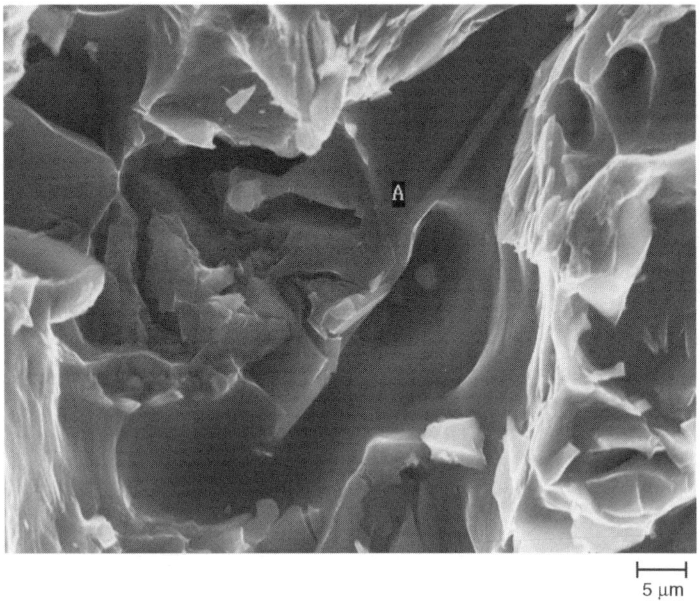

Fig. 5.59 Detail of [B] in Fig. 5.57. The needle-shape precipitate of the Al_7Cu_2Fe phase is situated in the deformed α-aluminum solid solution. 1500×

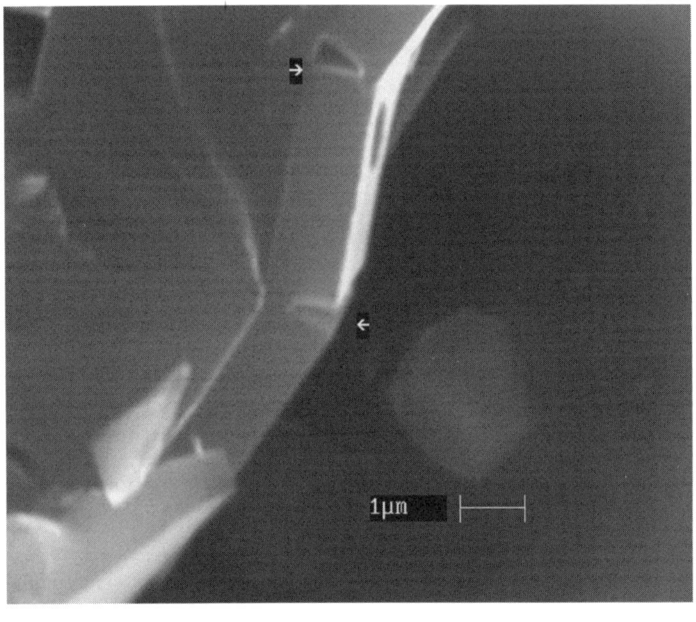

Fig. 5.60 Detail of [A] in Fig. 5.59. The microcracks, indicated by arrows, in the needle-shape precipitate of the Al_7Cu_2Fe phase are an effect of the local stress field. 10,000×

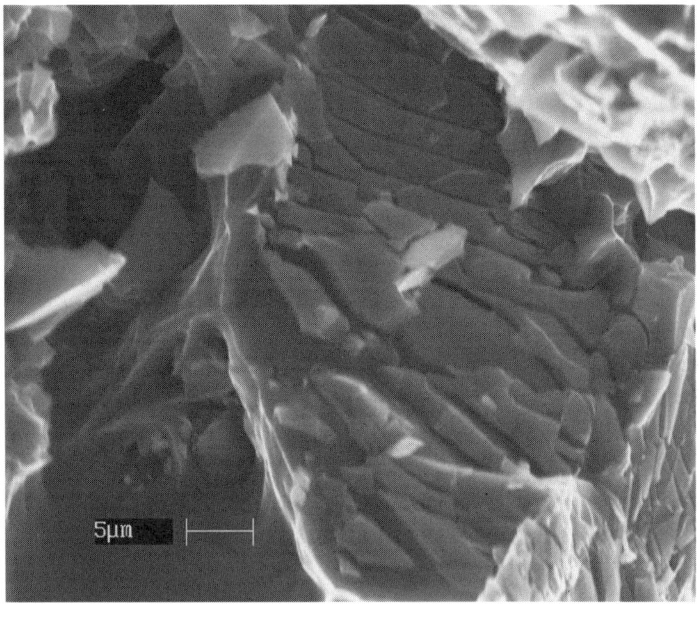

Fig. 5.61 Detail of [C] in Fig. 5.57. The crack in the silicon particles took place on the several cleavage planes. 2000×

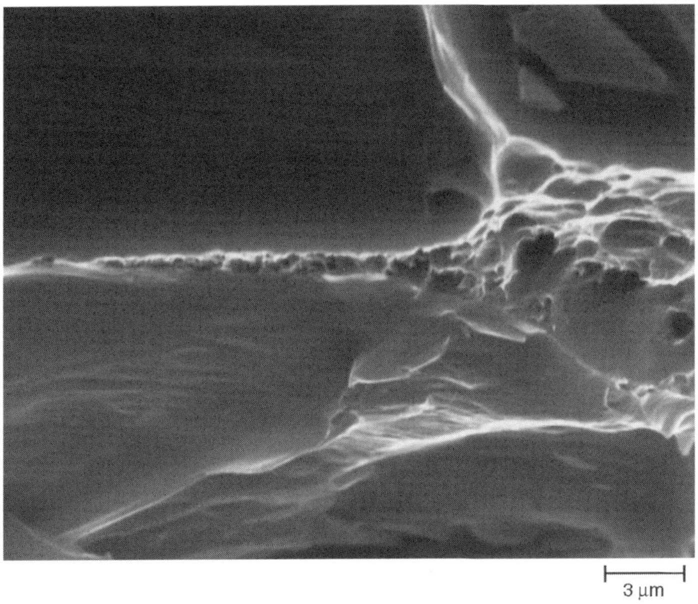

Fig. 5.62 Shear edge in the matrix zone. In two neighboring silicon precipitates, bands of waved steps of Wallner's lines morphology can be observed. 4000×

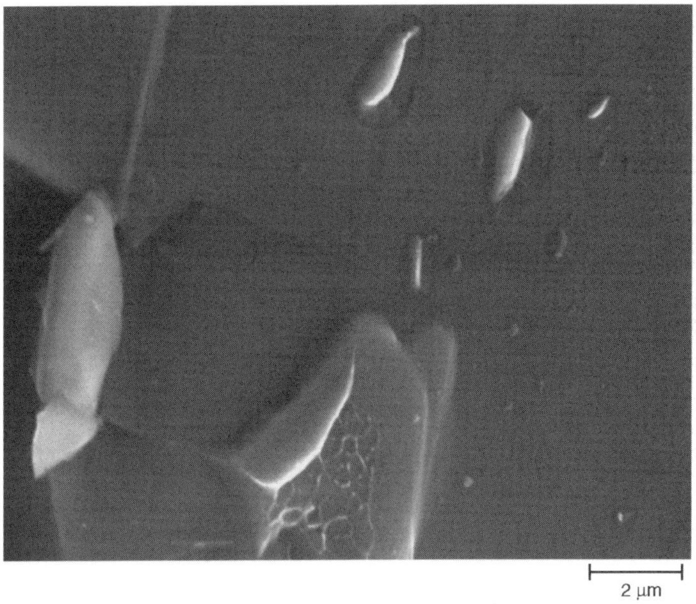

Fig. 5.63 Cleavage fracture morphology in the silicon particle. Steps and tongues on the cleavage facet are present. The characteristic uplift in the microdeformation boundary zone, where the interface cohesion was retained. 6500×

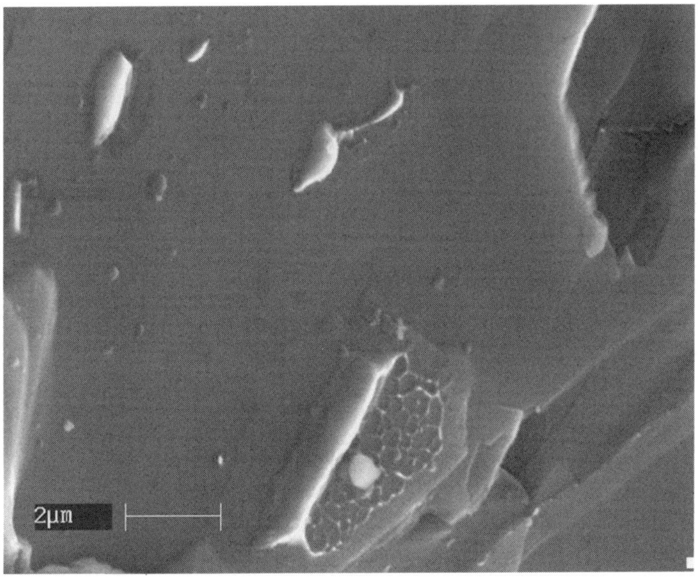

Fig. 5.64 Cleavage fracture morphology in the silicon particle. Steps and tongues on the cleavage facet are present. The characteristic uplift in the microdeformation boundary zone, where the interface cohesion was retained. 6500×

CHAPTER 6

Alloy 356.0 (AlSi7Mg)

Composition and Properties

Alloy	Designation	Chemical composition (main components), wt%						Properties(a)	
		Si	Cu	Mg	Mn	Ni	Fe	R_m, MPa	A_5, %
As-examined	AlSi7Mg	6.0–8.0	. . .	0.25–0.4	0.1–0.5	. . .	0.9 (max)	210	2
Aluminum Association standard(b)	356.0	6.5–7.5	0.25 max	02–0.45	0.35 max	. . .	0.6 max	228	5

(a) R_m, ultimate tensile strength; A_5, elongation measured over a length of 5.65 S_o, where S_o is the cross-sectional area of the test specimen before the test. (b) Alloy 356.0 is registered with the Aluminum Association and designated by ASTM for sand and permanent mold casting. R_m for 356.0-T6 in the minimum ultimate tensile strength for permanent mold casting.

Microstructures

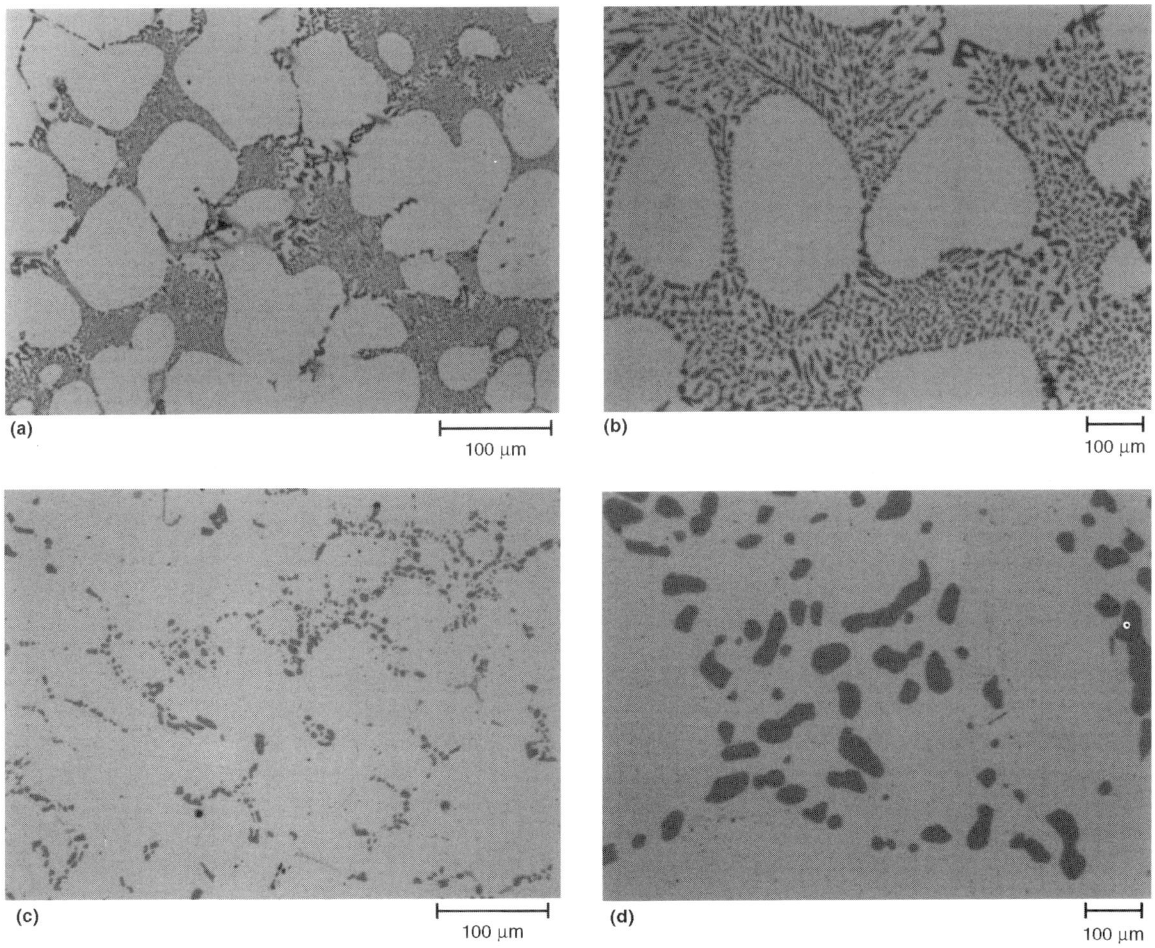

Fig. 6.1 Microstructures of AlSi7Mg (Alloy 356.0). Light microscope micrographs; etched with 1% HF. (a) As-cast modified, 150×. (b) As-cast modified, 750×. (c) After heat treatment (T6), 150×. (d) After heat treatment (T6), 750×

Fracture Profiles of Alloy 356.0 (AlSi7Mg), Refined, Modified, T6, Permanent Mold Casting

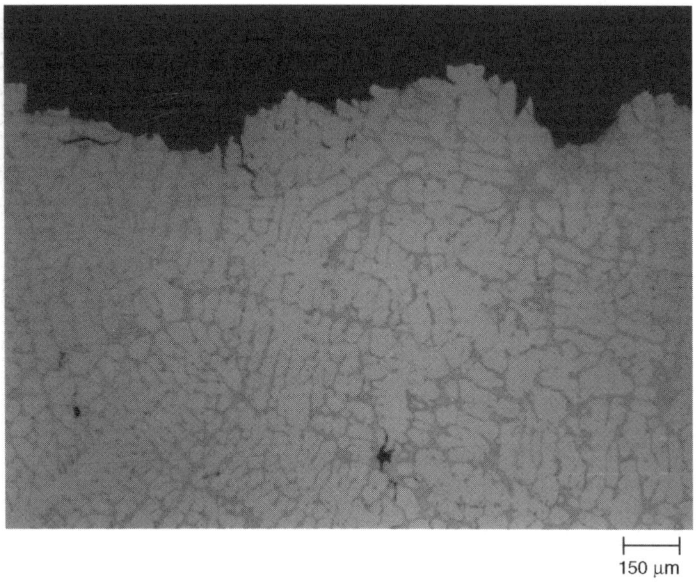

Fig. 6.2 Fracture profile of a specimen after static tensile test. The main crack crossed an interdendritic eutectic region. Secondary cracks are visible, as well as internal ones. They are parallel to the main crack profile. 50×

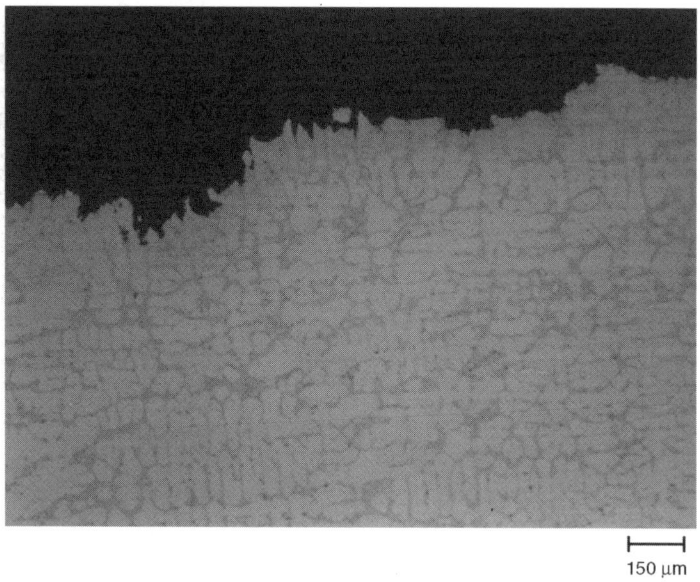

Fig. 6.3 Fracture profile of specimen after V-notch impact test at room temperature. The main crack crossed the interdendritic eutectic. Zigzag parts of the profile line are visible, formed by the edges of the shear ligaments in dendrites of the α-aluminum solid solution. 50×

Fig. 6.4 Fracture profile of specimen after V-notch impact test at –160 °C (–256 °F). The main crack crossed the region of the interdendritic eutectic. The zigzag elements of the profile line, formed by the edges of the shear micronecks of the α-aluminum solid solution in dendrite arms, are visible. Either secondary or internal cracks are visible. 200×

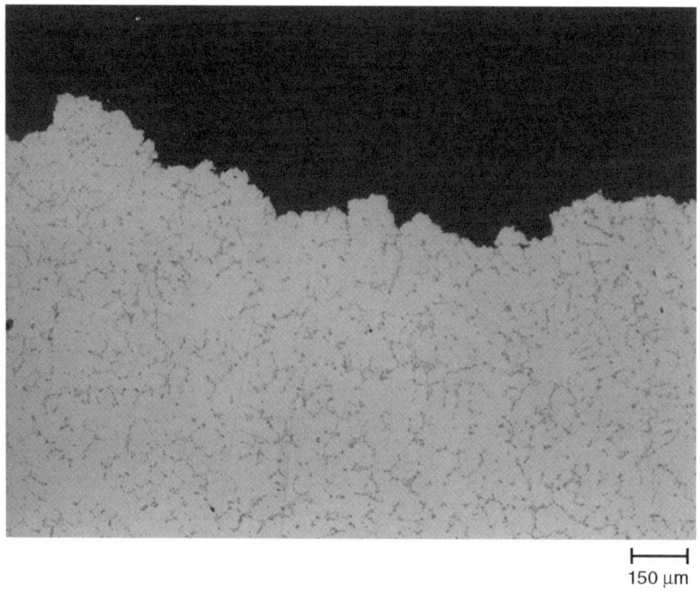

Fig. 6.5 Fracture profile of a specimen after static tensile test. The main crack line reflects the primary dendritic structure profile. 50×

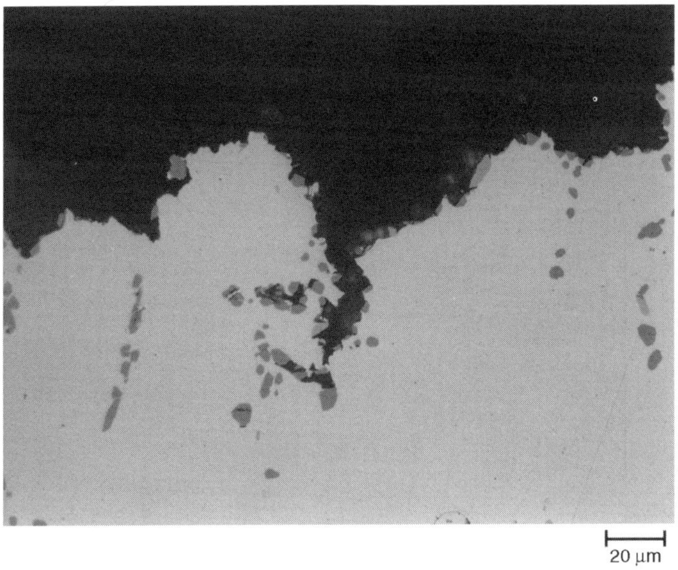

Fig. 6.6 Fracture profile of specimen after static tensile test. The main crack crossed two-phase regions. Cleavage cracks in silicon precipitates are visible. The sharp ends of the ligaments in α-aluminum solid solution and the secondary cracks in the interdendritic eutectic can be observed. Secondary cracks are visible in silicon precipitates and in brittle intermetallic phases. 400×

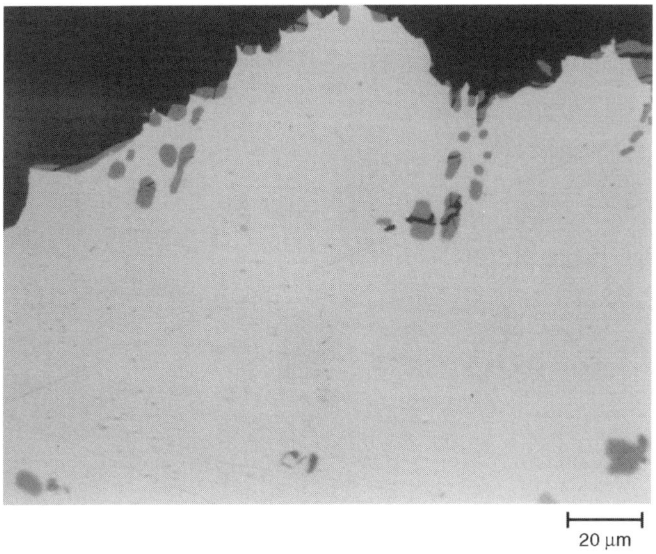

Fig. 6.7 Fracture profile of specimen after static tensile test. The main crack profile reflects the morphology of the two-phase regions. Cleavage cracks are visible in silicon precipitates on the profile line of the main crack. The sharp ends of the ligaments have formed in α-aluminum solid solution. Secondary cracks are observed in precipitates of silicon and brittle intermetallic phases. 500×

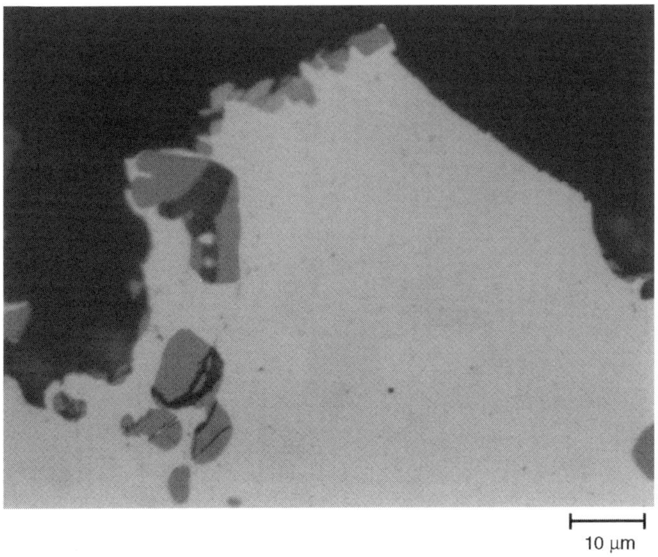

Fig. 6.8 Fracture profile of specimen after static tensile test. The main crack crossed two-phase regions (left part of micrograph). Cleavage cracks are visible in silicon precipitates in the profile line of the main crack. On the line of shear edge, in α-aluminum solid solution, the traces of the slip bands are visible, as a result of decohesion on successive slip planes. 1000×

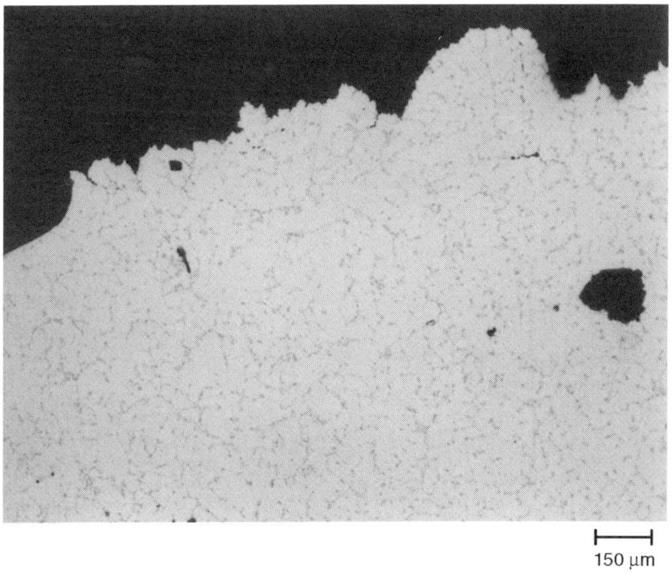

Fig. 6.9 Fracture profile of specimen after V-notch impact test at room temperature. The main crack crossed the interdendritic eutectic (rim zone of the specimen). 50×

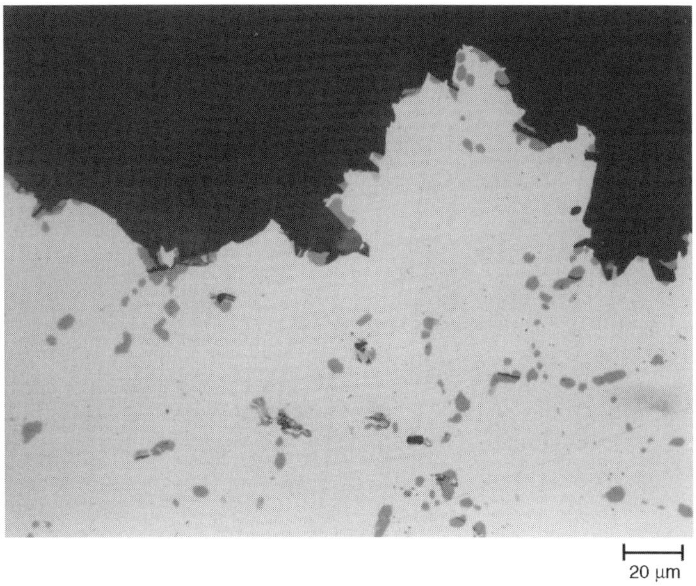

Fig. 6.10 Detail of the profile visible in Fig. 6.9. The main crack crossed two-phase regions. Cleavage cracks in silicon precipitates on the profile line of the main crack and the cleavage, internal cracks are visible. In α-aluminum solid solution the sharp ends of the ligaments after their plastic deformation have formed. The shear edges are situated in monophase regions in α-aluminum solid solution. 400×

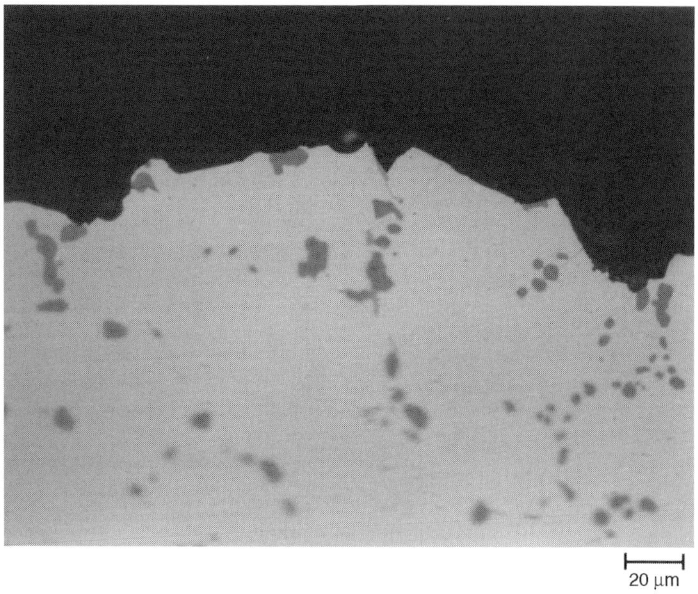

Fig. 6.11 Fracture profile of specimen after V-notch impact test at room temperature. The main profile line is formed by the shear edges in the monophase region of α-aluminum solid solution and by the cleavage lines in cracked silicon precipitates. 400×

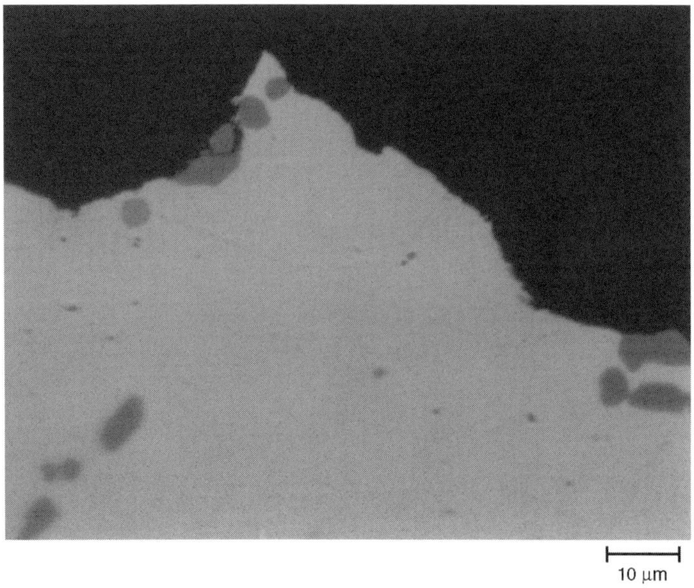

Fig. 6.12 Fracture profile of specimen after V-notch impact test at room temperature showing the sharp end of the deformed microneck of α-aluminum solid solution. Right side of the micrograph, shear edge with visible steps; left side, cleavage cracks in silicon particles. 1000×

Fracture Surfaces of Alloy 356.0 (AlSi7Mg), Refined, Modified, Permanent Mold Casting, Fracture after Static Tensile Test

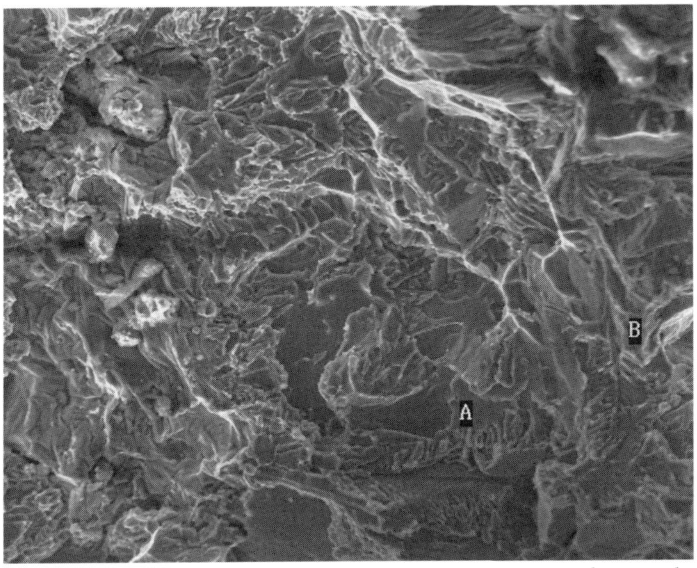

Fig. 6.13 Transcrystalline fracture of medium-developed surface, the large regions of the cleavage facets are visible in the silicon precipitates and the brittle intermetallic phases. The nonmetallic inclusions and secondary cracks also are shown. 400×

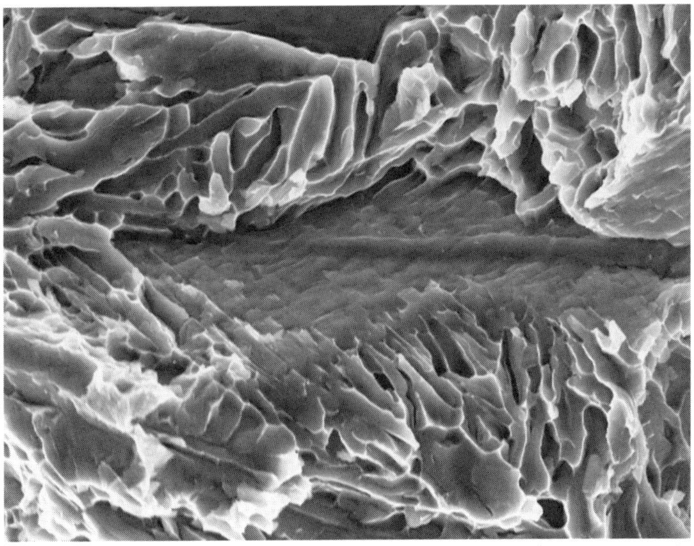

Fig. 6.14 Detail of [A] in Fig. 6.13. Transcrystalline fracture in the α-Al-(FeMn)Si phase precipitate. In α-aluminum solid solution both round precipitates and oval, open dimples are present, as a result of the material deformation. 2000×

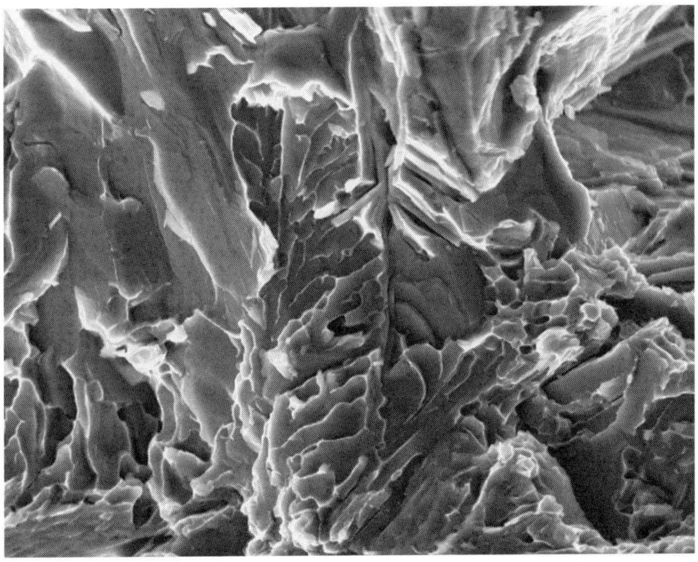

Fig. 6.15 Detail of [B] in Fig. 6.13. Transcrystalline fracture formed in the interdendritic eutectic of cleavage morphology in silicon particles and of cellular morphology in the two-phase region α-Al + α-Al(FeMn)Si. Rectilinear, secondary cracks also are visible. 1500×

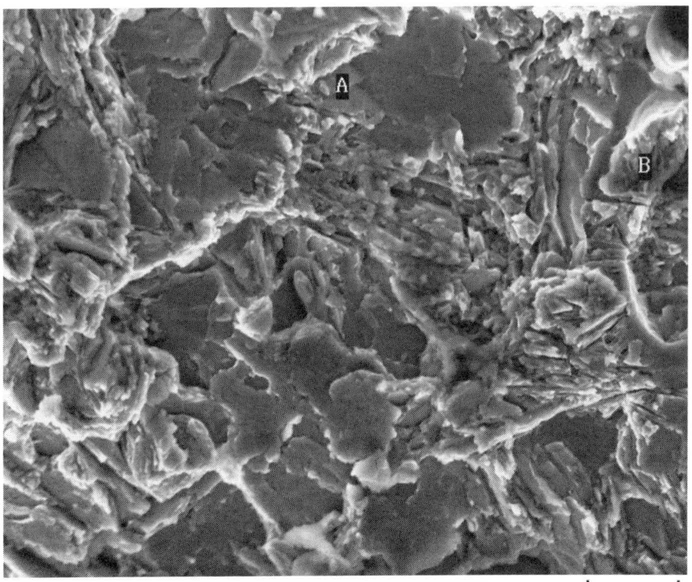

Fig. 6.16 Transcrystalline fracture of weakly developed surface. The main crack crossed the cleavage planes in the silicon precipitates, parallel to an average fracture plane in this microregion. The network of the rectilinear secondary cracks is visible. 700×

Fig. 6.17 Detail of [A] in Fig. 6.16. Cleavage facet with branched steps in eutectic silicon precipitate. 2000×

Fig. 6.18 Detail of [B] in Fig. 6.16. The tear ridge of micronecks in α-aluminum solid solution separates the two-phase regions of cellular morphology. 2000×

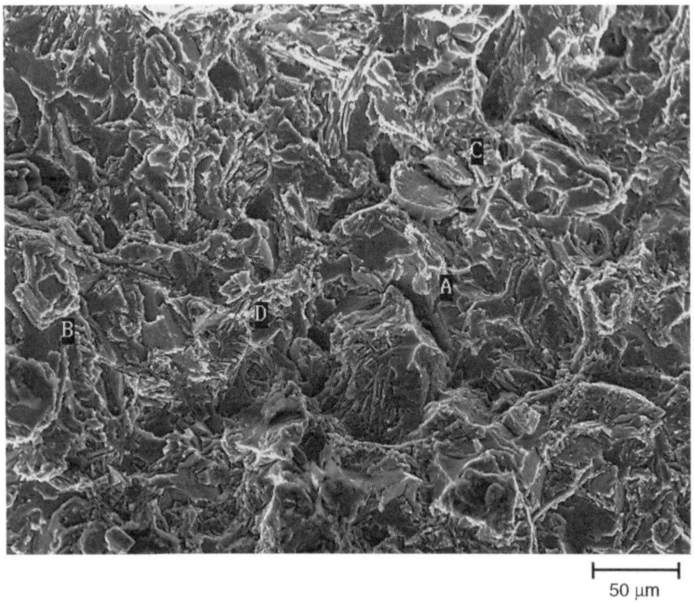

Fig. 6.19 Transcrystalline brittle fracture with the surface greatly developed. Cleavage facets and secondary cracks are visible. 260×

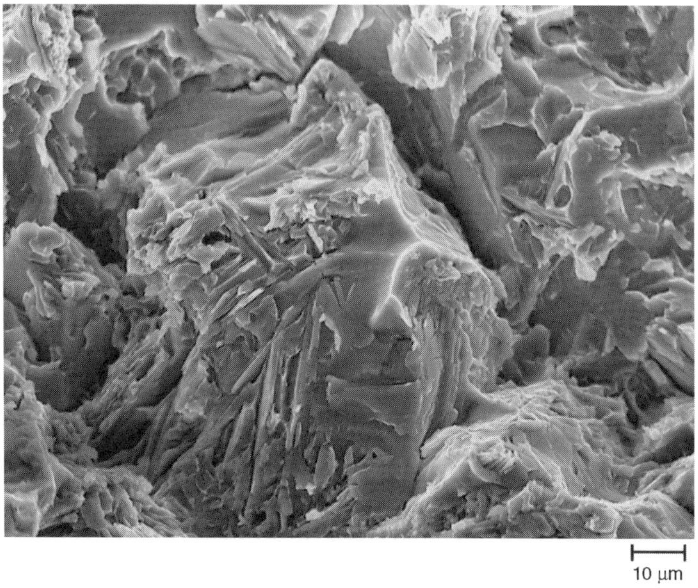

Fig. 6.20 Detail of [A] in Fig. 6.19. On the fracture surface, a crack in the intermetallic α-Al(FeMn)Si phase is visible. The network of steps was formed during the crack front displacement on the successive cleavage planes. A secondary crack is visible on the interface. 800×

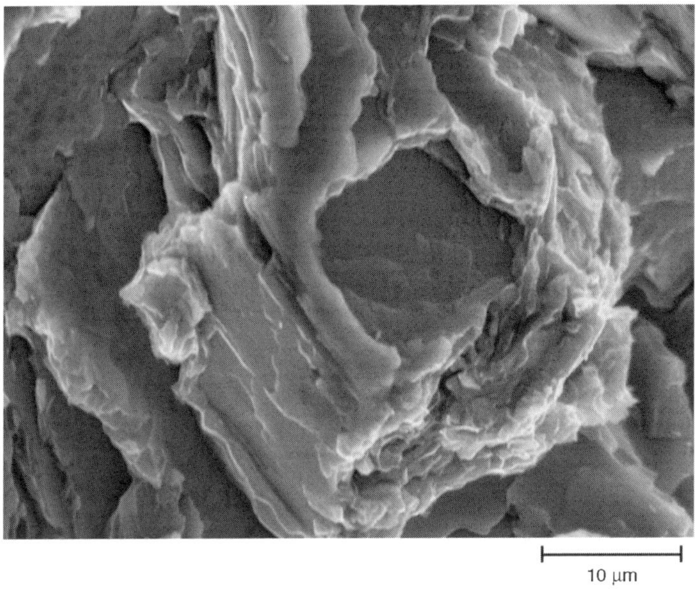

Fig. 6.21 Detail of [B] in Fig. 6.19. Cleavage steps among the parallel cleavage planes in the eutectic silicon precipitates. 2000×

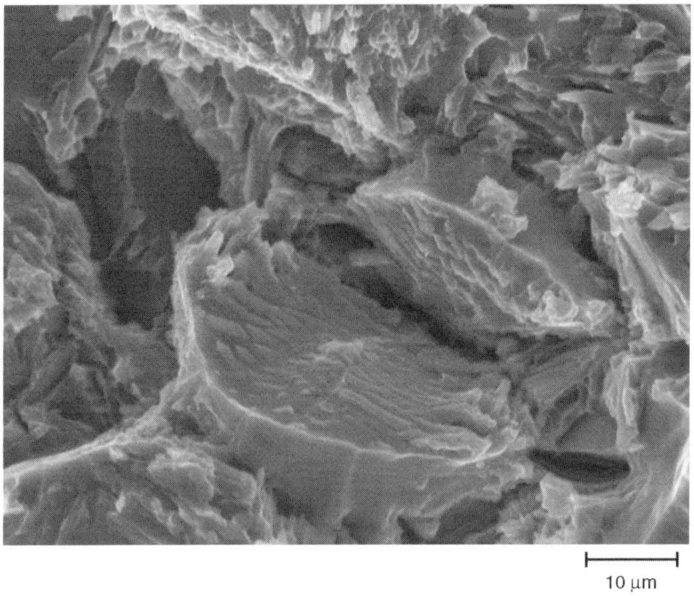

Fig. 6.22 Detail of [C] in Fig. 6.19. Secondary crack formed in the intermetallic phase particle Mg_2Si. The developed system of steps is a result of the multiplane cracks. The microneck of the α-aluminum solid solution fractured among brittle phase particles. 1300×

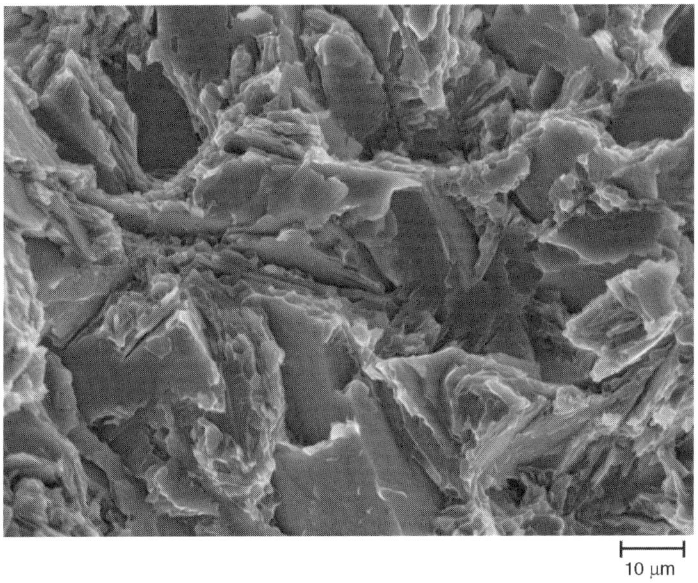

Fig. 6.23 Detail of [D] in Fig. 6.19. Transcrystalline, cleavage fracture in the eutectic microregion: α-Al + Si + β-AlFeSi. Rectilinear secondary cracks are visible. In the β-AlFeSi particle, the crack displaced on the several cleavage planes of different orientations. 800×

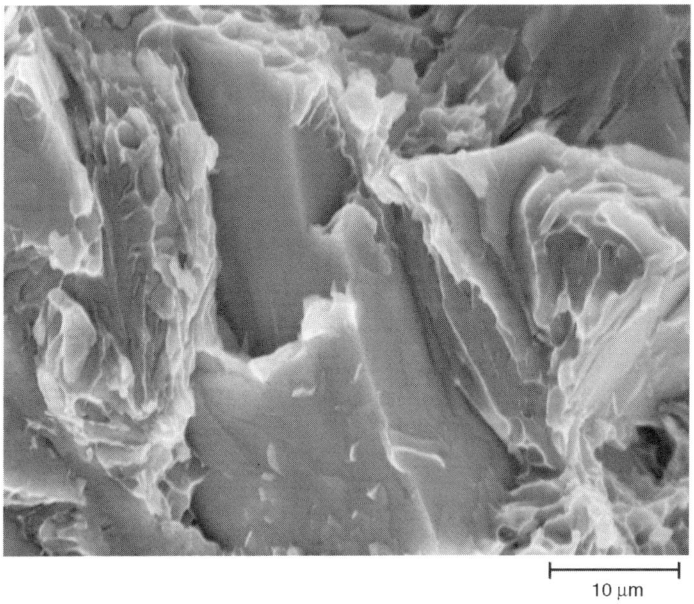

Fig. 6.24 Detail of Fig. 6.23. On the fracture surface, the particle of the brittle β-AlFeSi phase is visible. The steps of different height were formed on the cleavage facets. The morphology of oval and open dimples in the matrix is characteristic for the shear process. 1700×

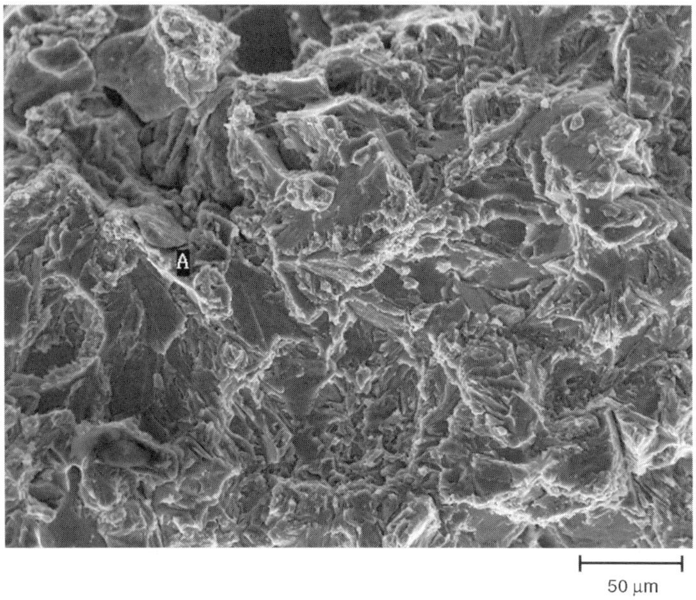

Fig. 6.25 Transcrystalline brittle fracture of medium-developed surface. The main crack was formed in several cleavage planes in the eutectic silicon precipitates. The particles of the intermetallic phase are visible. 300×

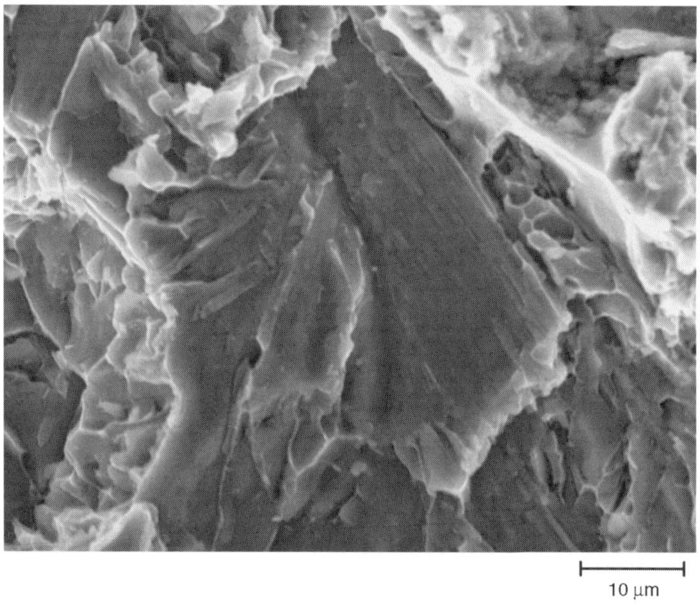

Fig. 6.26 Detail of [A] in Fig. 6.25. On the surface of the cracked particle of eutectic silicon, the bands of the parallel steps were formed. In α-aluminum solid solution, near to the interface, the effects of the microdeformation are visible. 1500×

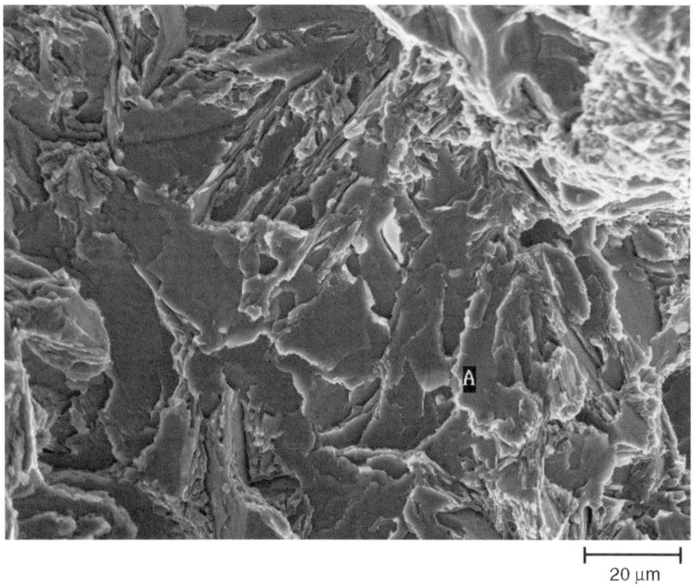

Fig. 6.27 Transcrystalline fracture of weakly developed surface. The cleavage crack crossed the eutectic silicon precipitates; the cleavage steps are shown among the parallel cleavage planes. The visible secondary cracks are rectilinear. 700×

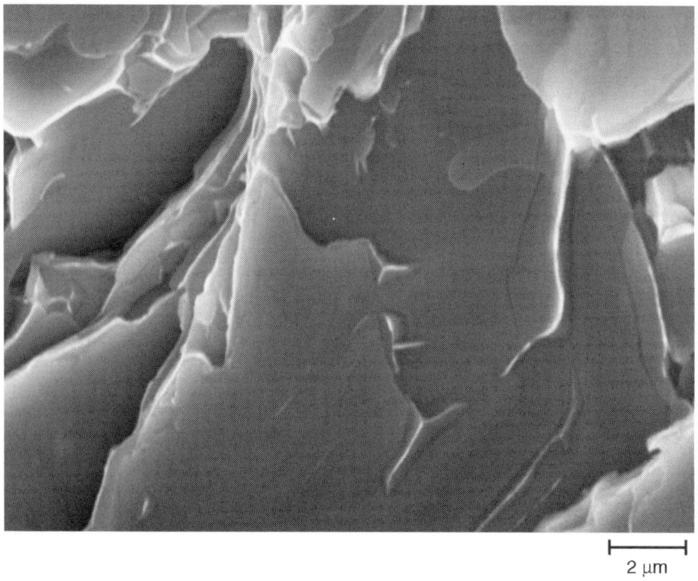

Fig. 6.28 Detail of [A] in Fig. 6.27. The system of the branched cleavage steps was revealed on the cleavage planes. Tongues formed in the regions of microdeformation (crystal defects) in the silicon crystal. 5500×

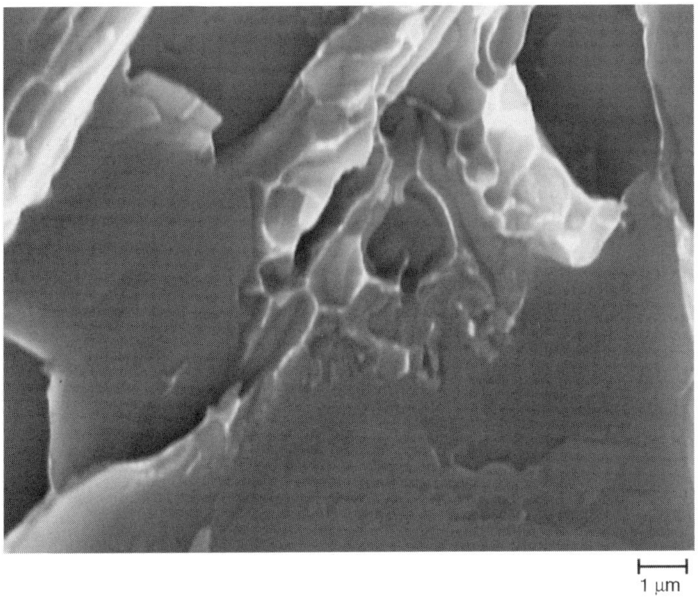

Fig. 6.29 Transcrystalline cleavage fracture. Among the eutectic silicon plates, on the interface between α-aluminum and silicon, the regions of well-retained cohesion are visible. The results of plastic microdeformation are revealed. The small cleavage steps are present on the cleavage facets in the silicon. 7000×

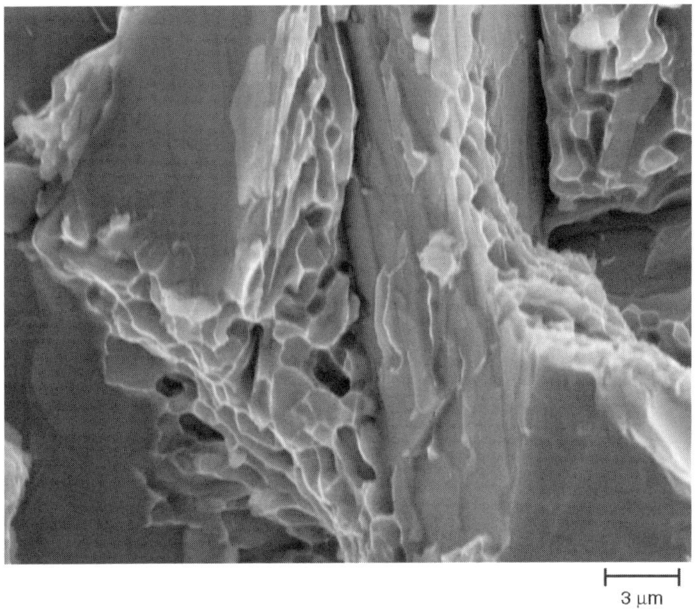

Fig. 6.30 Fracture of mixed morphology. In α-aluminum solid solution the dimples were formed as a result of plastic deformation. The crack has propagated in the silicon particles on the cleavage planes. 3500×

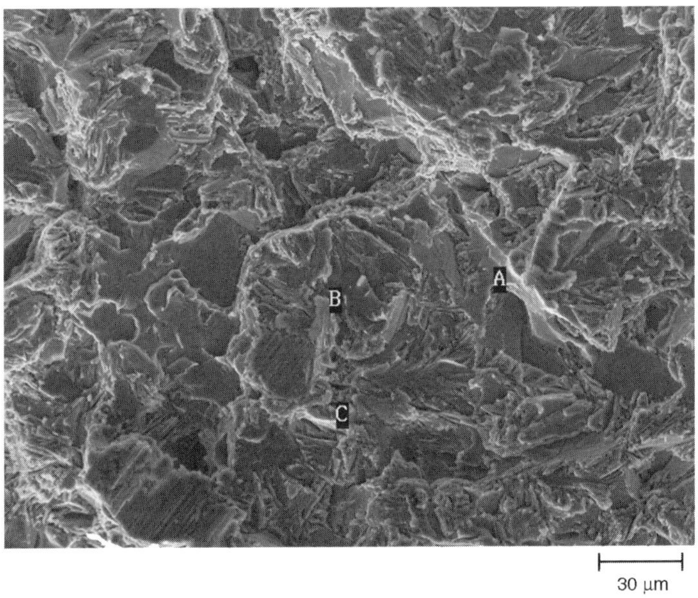

Fig. 6.31 Transcrystalline fracture of medium-developed surface. The regions of cleavage morphology in silicon particles and of cellular morphology in two-phase zone are visible. 400×

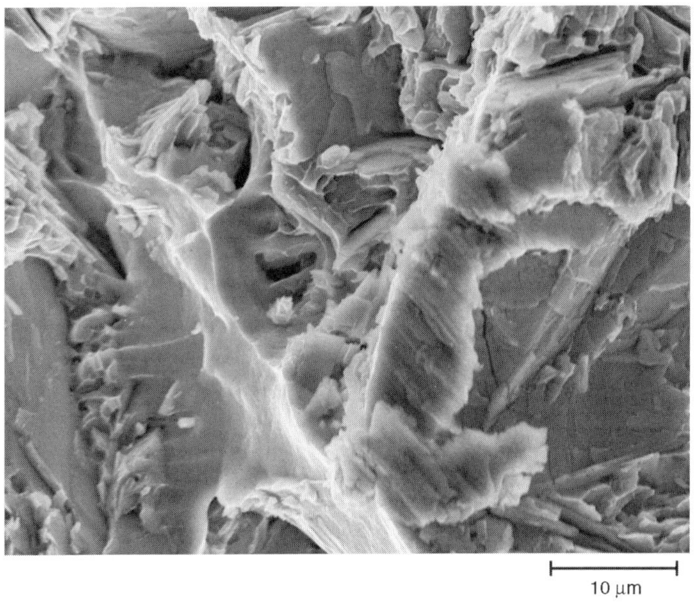

Fig. 6.32 Detail of Fig. [A] in 6.31. The tear ridges and the micronecks in α-aluminum solid solution are visible. Traces of plastic microdeformation and parallel slip bands were revealed on the shear surface. 1800×

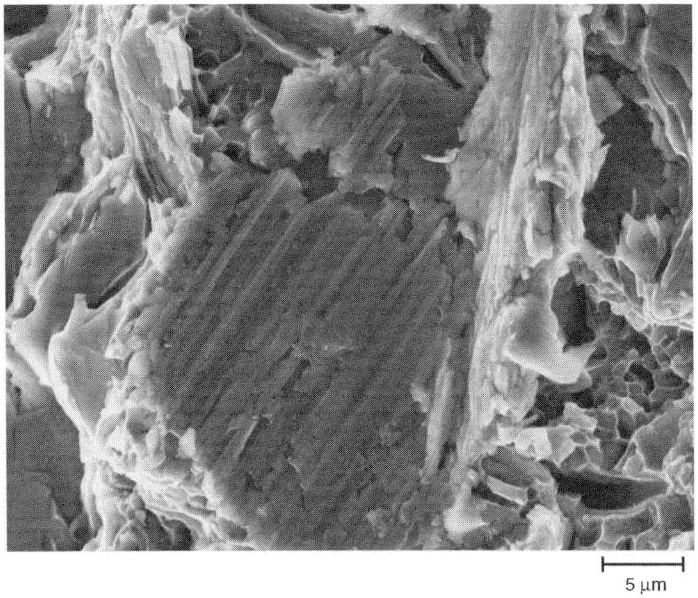

Fig. 6.33 Detail of [B] in Fig. 6.31. In α-aluminum solid solution, the slip bands, secondary cracks, and dimples characteristic of plastic deformation are shown. 2400×

Fig. 6.34 Detail of Fig. 6.33. Slip bands are visible on the shear surface in α-aluminum solid solution. 7500×

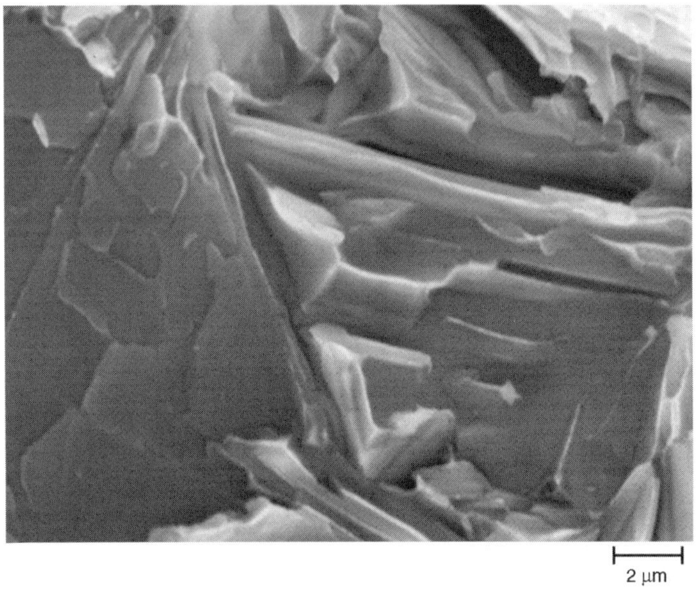

Fig. 6.35 Detail of [C] in Fig. 6.31. The network of the rectilinear steps of different height, formed on the surface of the cracked Mg_2Si particle. Secondary cracks also are visible. 5000×

Fracture Surfaces of Alloy 356.0 (AlSi7Mg), Refined, Modified, Permanent Mold Casting, Fracture after V-Notch Impact Test at 21 °C (70 °F)

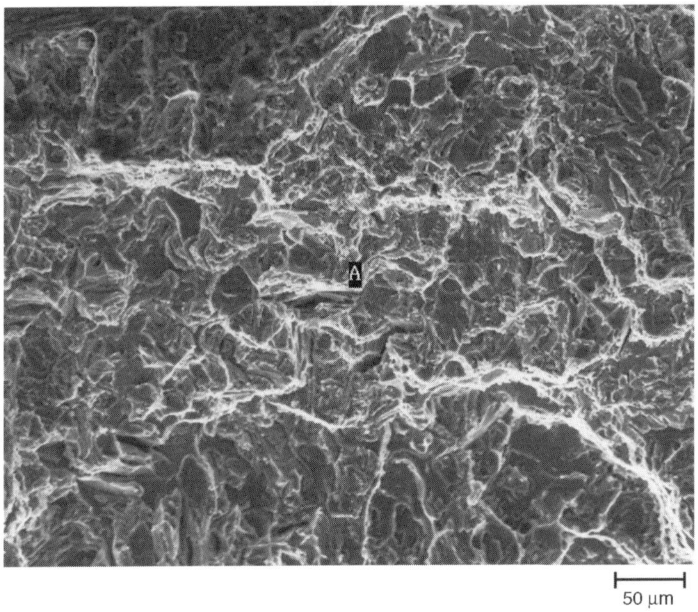

Fig. 6.36 Transcrystalline fracture of greatly developed surface and the morphology characteristic of cleavage fracture. 200×

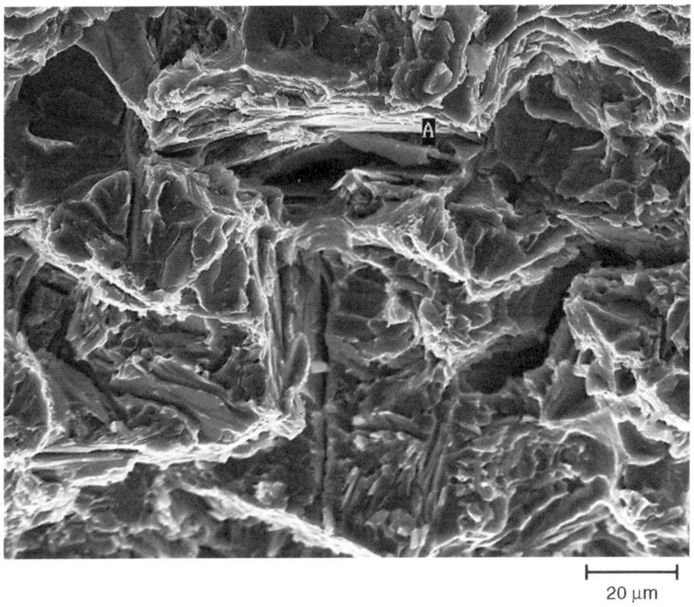

Fig. 6.37 Detail of [A] in Fig. 6.36. Zigzag bands of the rectilinear steps on the surface transcrystalline fracture and the network of the secondary cracks are visible. 650×

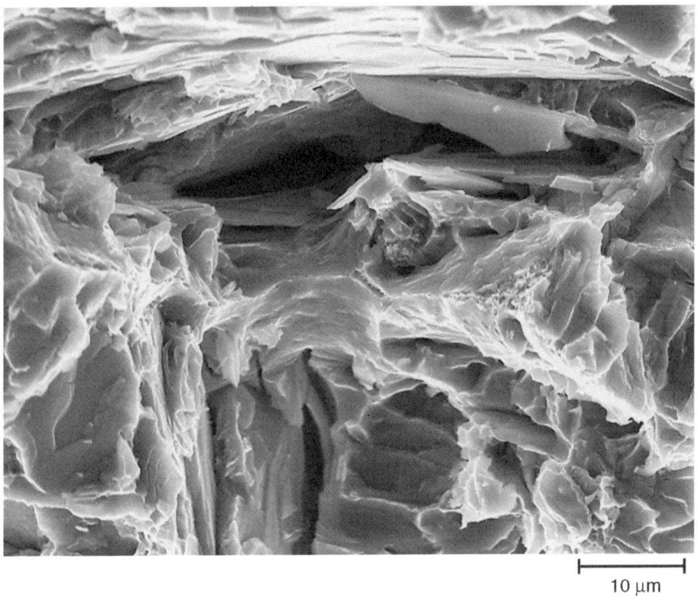

Fig. 6.38 Detail of [A] in Fig. 6.37. The edges of the deformed and fractured micronecks in α-aluminum solid solution with visible traces of the microdeformation (dimples) are shown in the crack zone. The branched bands of the secondary cracks and of the cleavage steps formed on the cleavage facets in the silicon particles. 1500×

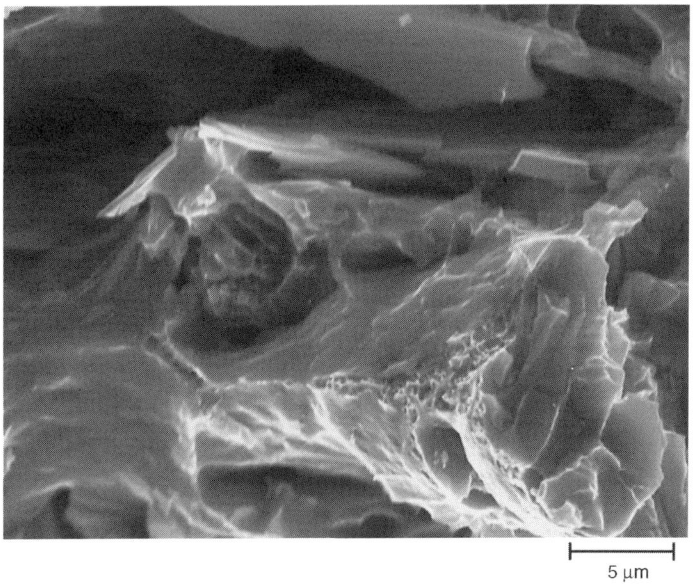

Fig. 6.39 Detail of Fig. 6.38. Dimples resulting from plastic deformation formed in the crack zone of the microligaments of α-aluminum solid solution. 3000×

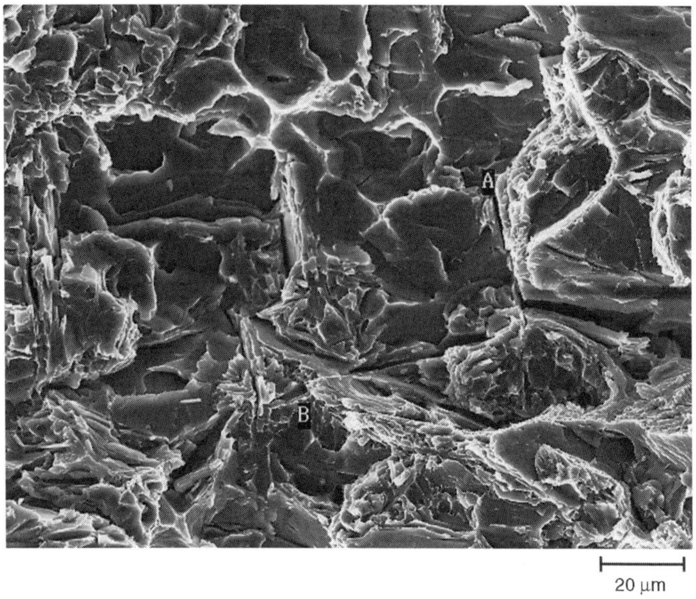

Fig. 6.40 The large transcrystalline cleavage cracks are visible in the eutectic silicon particles. (The crack front crossed several cleavage planes.) Branched steps and secondary cracks also are visible. In the matrix, as a result its plastic deformation, the micronecks were formed. 600×

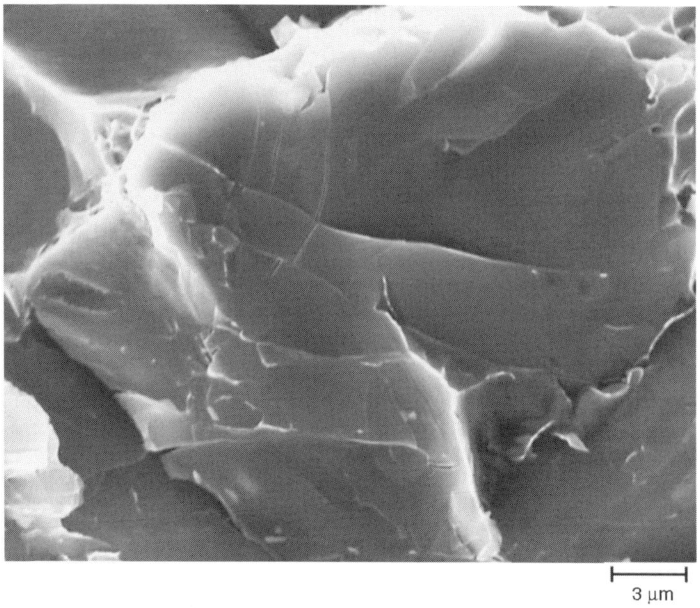

Fig. 6.41 Detail of [A] in Fig. 6.40. Morphology of the cleavage steps forming a rectangular network on the cleavage facets in the silicon precipitate is shown. Plastic microdeformation caused the dimple formation on the interface between α-aluminum and silicon. 3700×

Fig. 6.42 Detail of [B] in Fig. 6.40. Transcrystalline cleavage fracture of mixed morphology. In the silicon precipitates, the crack propagated on the cleavage planes. The dimples have formed during plastic microdeformation of the α-aluminum solid solution, situated among plates of the eutectic silicon. 1800×

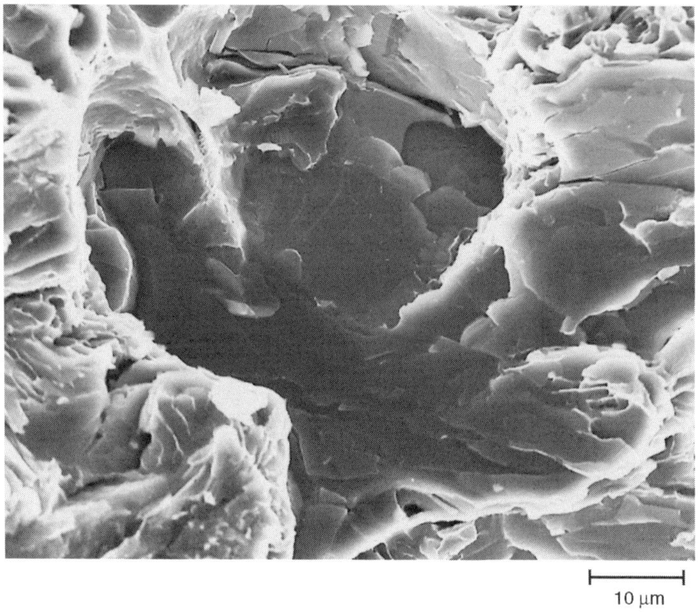

Fig. 6.43 Cleavage crack in the silicon precipitate. On the cleavage facets of the silicon crystals, the numerous cleavage steps are visible. The small dimples visible on the microneck edge in the α-aluminum solid solution are a result of its plastic deformation. 1300×

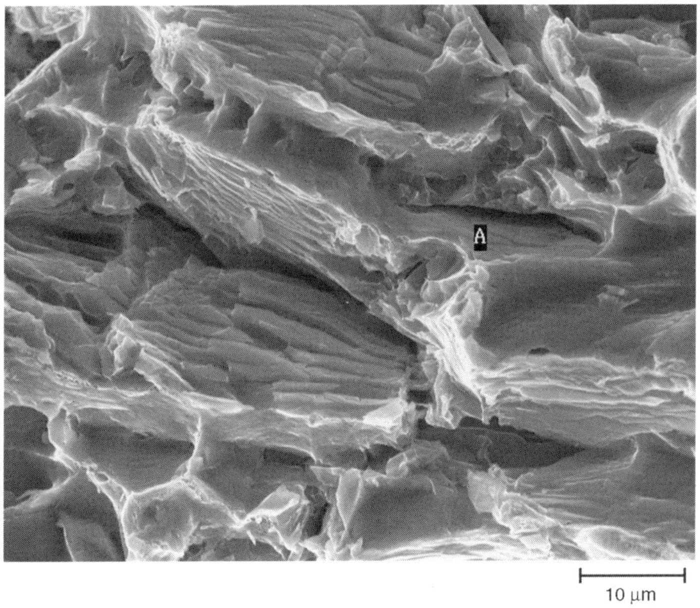

Fig. 6.44 Cleavage steps formed in the branched particles of the intermetallic phase α-Al(FeMn)Si. Numerous secondary cracks are visible in this brittle phase. The cleavage step system is a result of the crack front propagation on the successive cleavage planes. 1500×

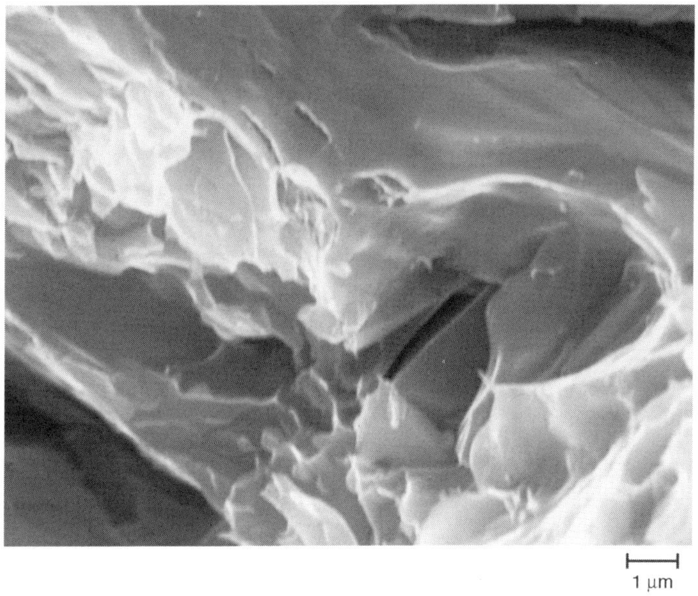

Fig. 6.45 Detail of [A] in Fig. 6.44. The system of the step shelves was formed during propagation of the crack front to the successive cleavage planes. 7000×

Fractures Surfaces of Alloy 356.0 (AlSi7Mg), Refined, Modified, Permanent Mold Casting, Fracture after V-Notch Impact Test at –160 °C (–256 °F)

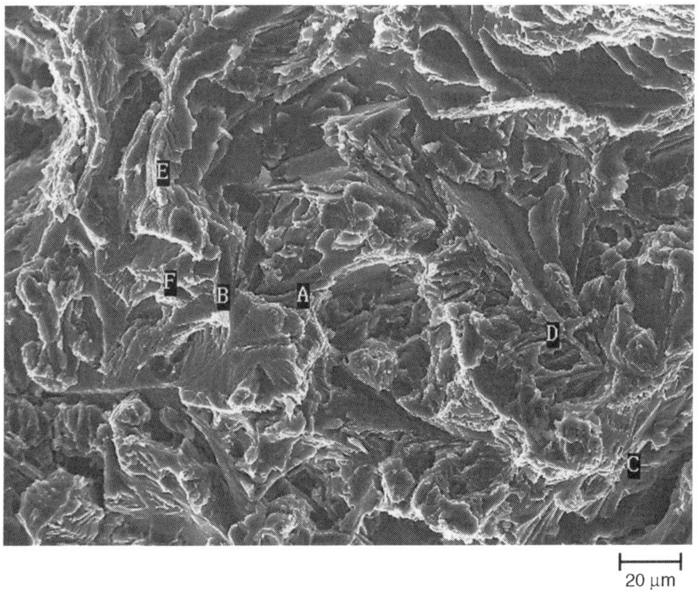

Fig. 6.46 Transcrystalline fracture with the greatly developed surface. The main crack crossed the silicon precipitate in the cleavage planes of different orientation in relation to the average fracture plane. The rectilinear secondary cracks, the branched cleavage steps, and the tear ridges have formed in the α-aluminum solid solution. 450×

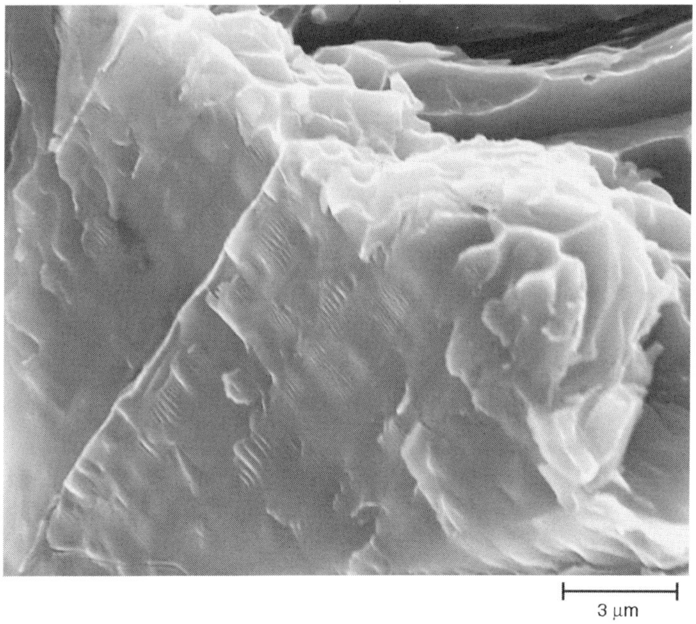

Fig. 6.47 Detail of [A] in Fig. 6.46. The surface of the cracked silicon particle is shown. On the cleavage facet between two parallel cleavage steps, bands of waved short steps and tongues in the microdeformation zones of the silicon crystal lattice are visible. 5500×

Fig. 6.48 Detail of [B] in Fig. 6.46.The bands of the parallel steps on the cleavage facets in eutectic silicon were revealed. On the microneck edge in the α-aluminum solid solution, between two silicon particles, the oval, open dimples were formed. 3300×

Fig. 6.49 Detail of [C] in Fig. 6.46. The crack zone of the α-aluminum solid solution is situated along the tear ridge of the microneck. Open dimples are visible on the shear surface. The bands of the waved, crossed steps, of Wallner's lines morphology, were formed on the cleavage facets in the silicon particles. 3000×

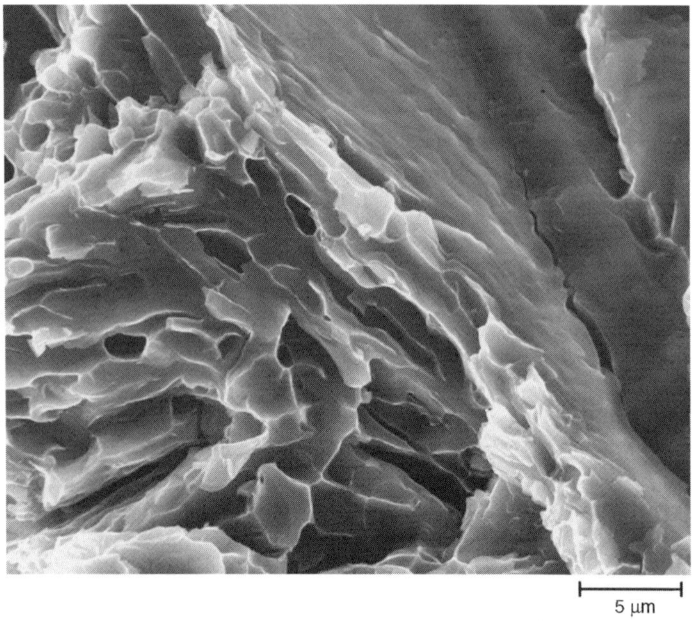

Fig. 6.50 Detail of [D] in Fig. 6.46. Fracture of mixed morphology. Oval shear dimples are situated near the cleavage facet. The secondary brittle crack took place on the interface between α-aluminum and silicon. 3000×

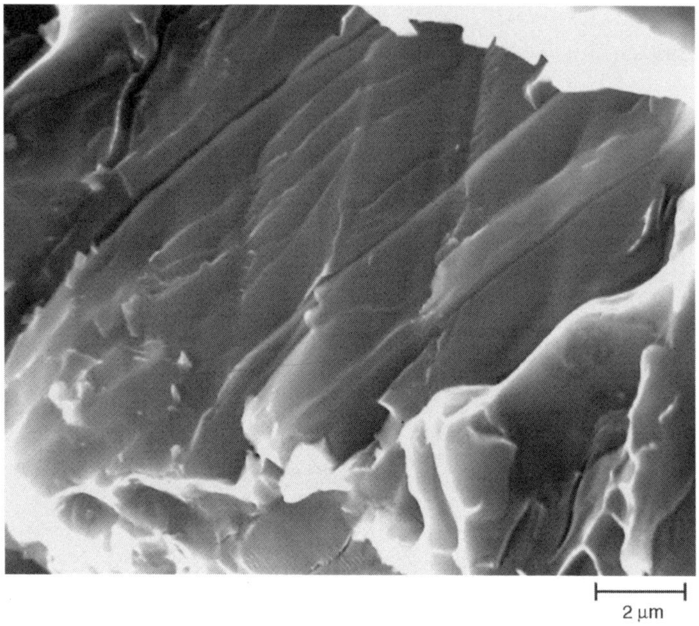

Fig. 6.51 Detail of [E] in Fig. 6.46. The cleavage step bands intersect at a 45° angle on the cleavage facet of the silicon particles. 5500×

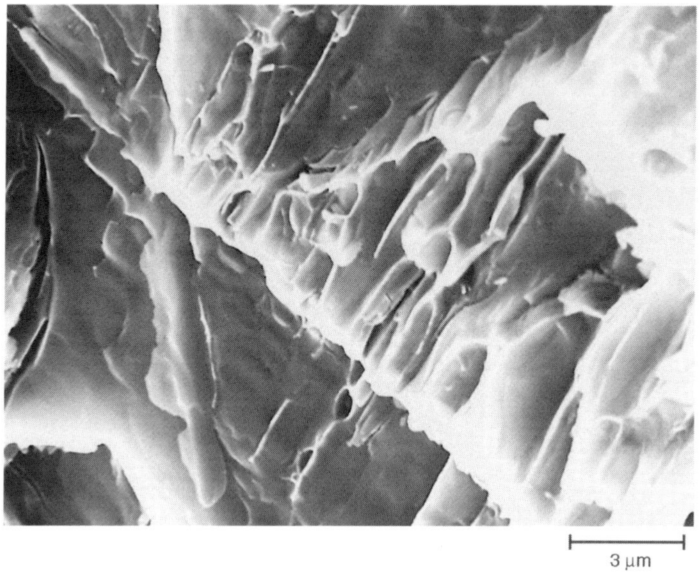

Fig. 6.52 Detail of [F] in Fig. 6.46. Oval, open, shear dimples formed in the shear matrix zone. 5500×

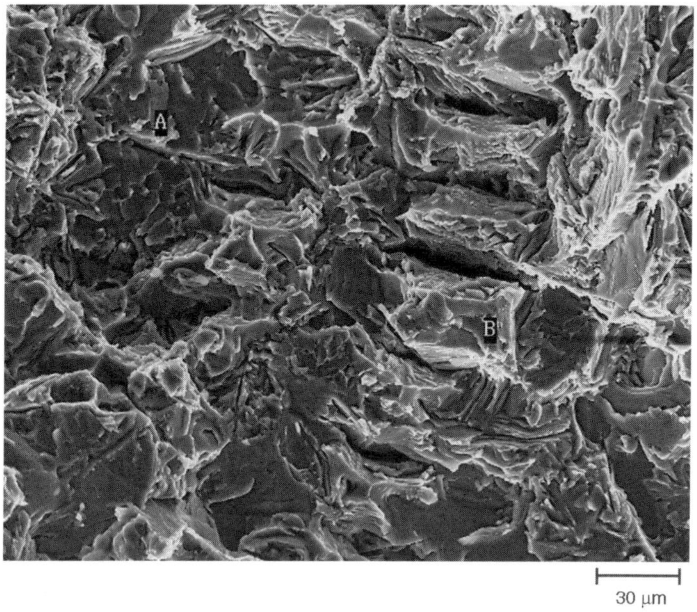

Fig. 6.53 Transcrystalline brittle fracture. The band of the rectilinear, parallel, secondary cracks is visible. 400×

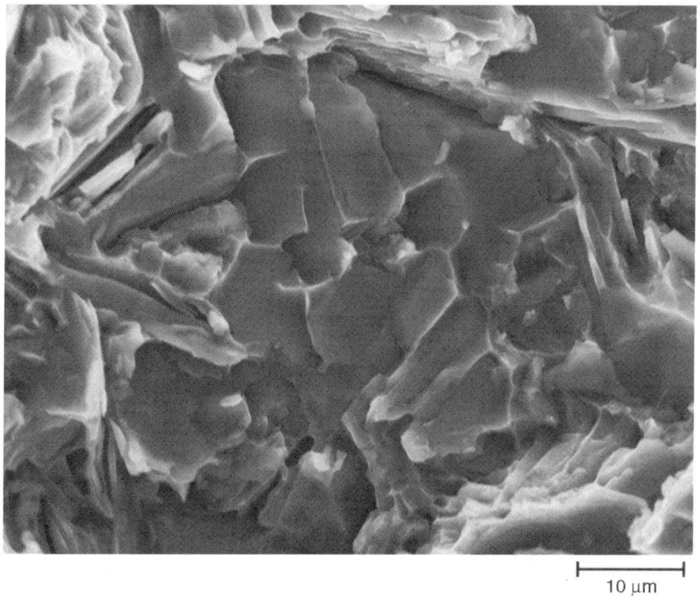

Fig. 6.54 Detail of [A] in Fig. 6.53. In the silicon particle the system of the cleavage steps is visible. The steps intersected at an angle between 45° and 90° on the cleavage facets of different orientation. Secondary cracks were formed on the interface between α-aluminum and silicon. 1550×

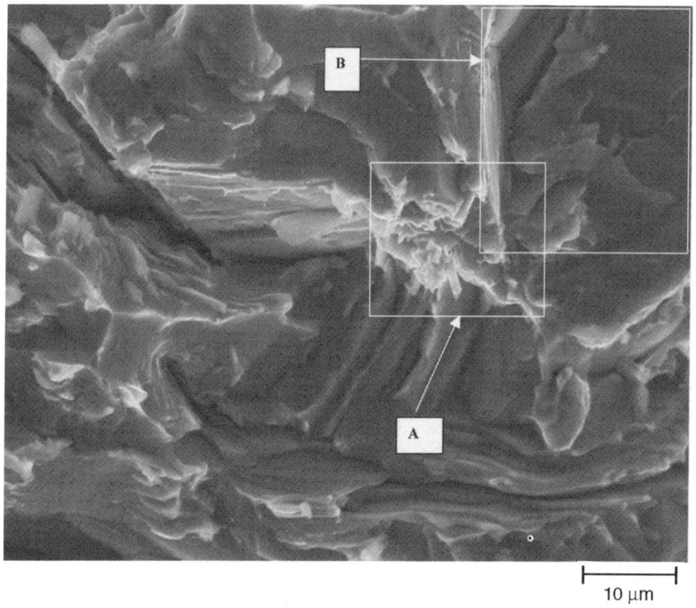

Fig. 6.55 Detail of [B] in Fig. 6.53. Fracture surface formed between two cleavage steps. The step bands and the secondary cracks, situated perpendicularly to the average fracture plane also are visible. 1400×

Fig. 6.56 Detail of [A] in Fig. 6.55. Surface of the silicon plate cracked several times. The crack front crossed the numerous cleavage planes of different orientation. 5000×

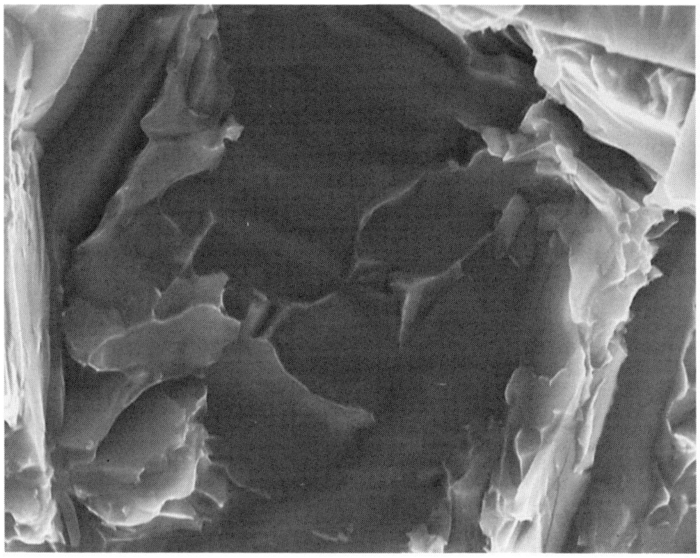

Fig. 6.57 Detail of [B] in Fig. 6.55. Branching of cleavage steps on the cleavage facets in the eutectic silicon plate. 2800×

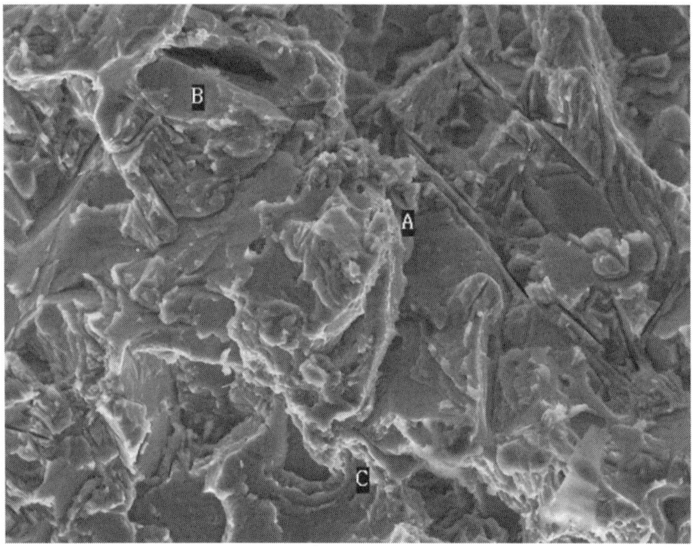

Fig. 6.58 Transcrystalline fracture of weakly developed surface. The network of the perpendicular secondary cracks was formed. Most of the visible cleavage facets in the brittle phases are parallel to the average fracture plane. 500×

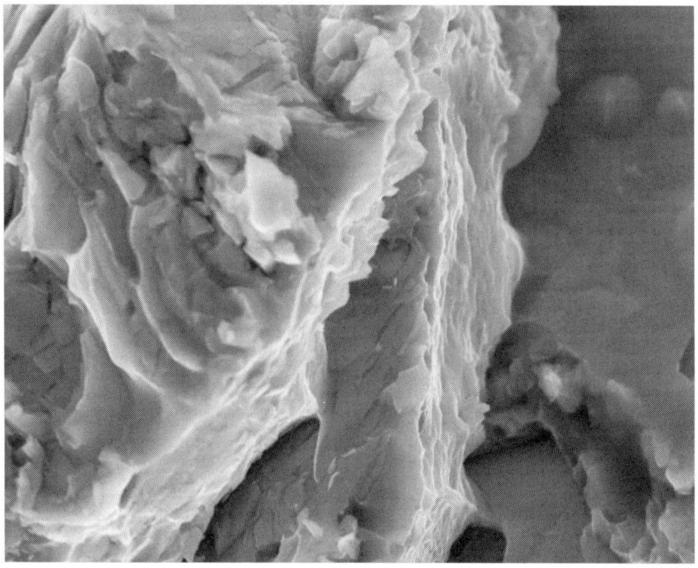

Fig. 6.59 Detail of [A] in Fig. 6.58. Traces of plastic deformation in the shape of the dimples on the tear ridge of the micronecks were revealed in the α-aluminum solid solution. 2600×

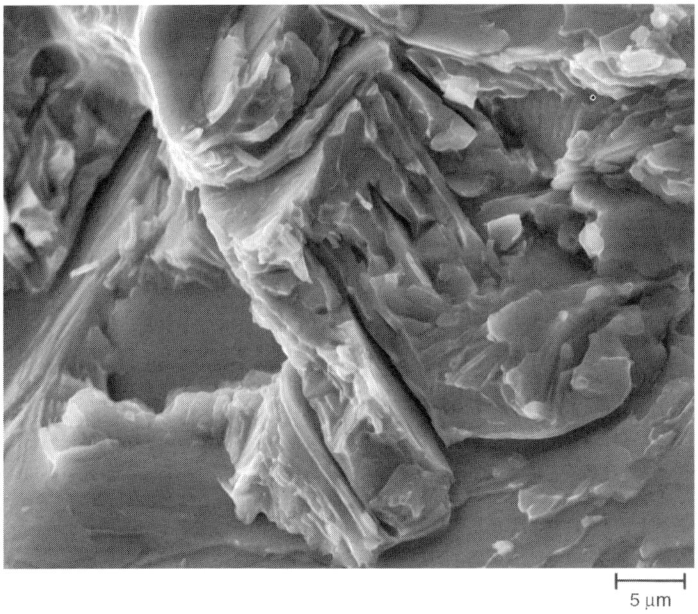

Fig. 6.60 Detail of [B] in Fig. 6.58. The chevron formed by secondary crack and step bands. In the left top corner of the micrograph, the cracked microneck of the α-aluminum solid solution is visible. 2000×

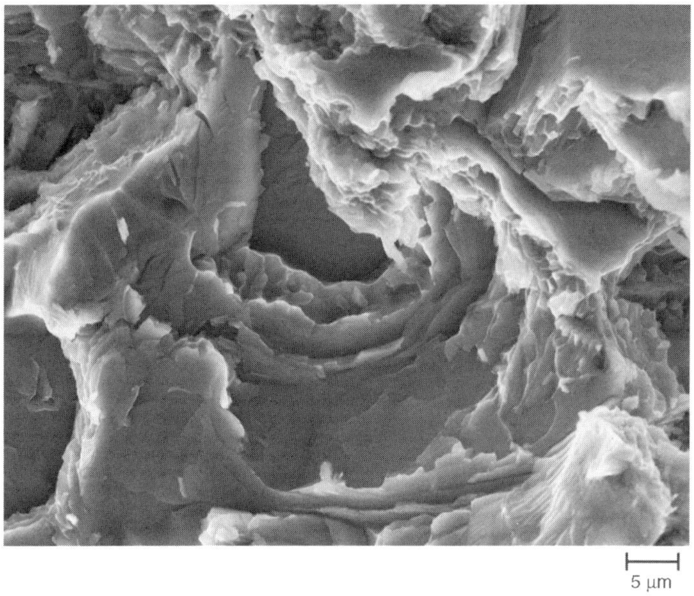

Fig. 6.61 Detail of [C] in Fig. 6.58. The eutectic silicon particle is surrounded with tear ridge of the microneck of the α-aluminum solid solution. The morphology of the dimples in the matrix is characteristic of the shear process. The terrace system of the cleavage steps resulted from multiplied brittle cracks in the silicon particle. 1500×

Fracture Surfaces of Alloy 356.0 (AlSi7Mg), Refined, Modified, Permanent Mold Casting, T6, Fracture after Static Tensile Test

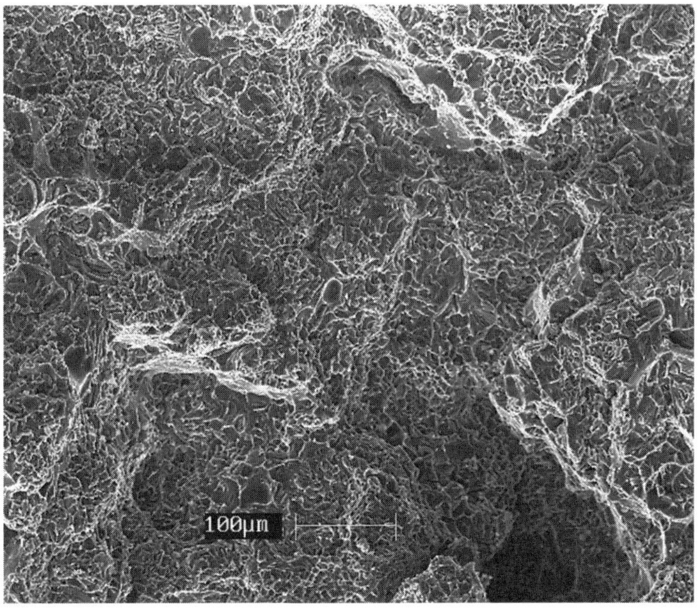

Fig. 6.62 Transcrystalline fracture of medium-developed surface. The areas of fracture of both cleavage and cellular features are visible (see Fig. 6.2). 150×

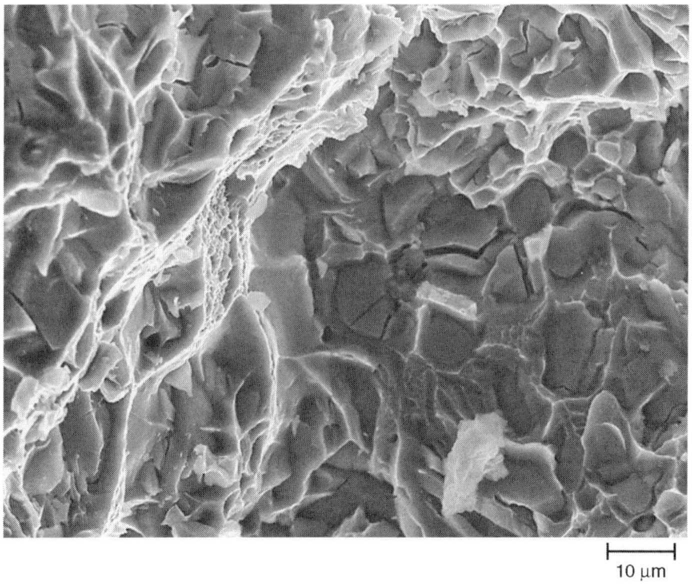

Fig. 6.63 Transcrystalline fracture of cellular morphology. The crack front crosses the cleavage planes in the silicon particles. The deformed micronecks of the α-aluminum solid solution are situated among them (see Fig. 6.7). On the edge of these micronecks (left part of the micrograph) the dimples are visible, which can be a result of the mixed mechanism of fracture in this region. Numerous secondary cracks are shown in the silicon particle. 1000×

Fig. 6.64 Morphology of the fracture surface in the vicinity of the deep secondary crack in two-phase region. Zones of well-retained cohesion on the interfaces between α-aluminum and silicon are visible (see Fig. 6.6). 1900×

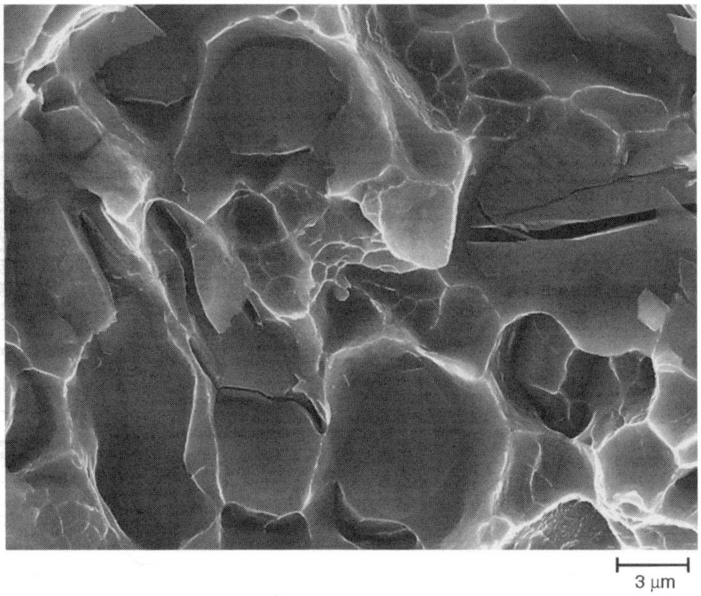

Fig. 6.65 Fracture in the two-phase region. The early stages of decohesion are visible on the interfaces between α-aluminum and silicon. In most of the silicon particles the cleavage facets are situated parallel (see Fig. 6.8). In the microregions of the α-aluminum solid solution, the dimples have formed around the cracked silicon particles, as a result of plastic deformation of matrix. 4000×

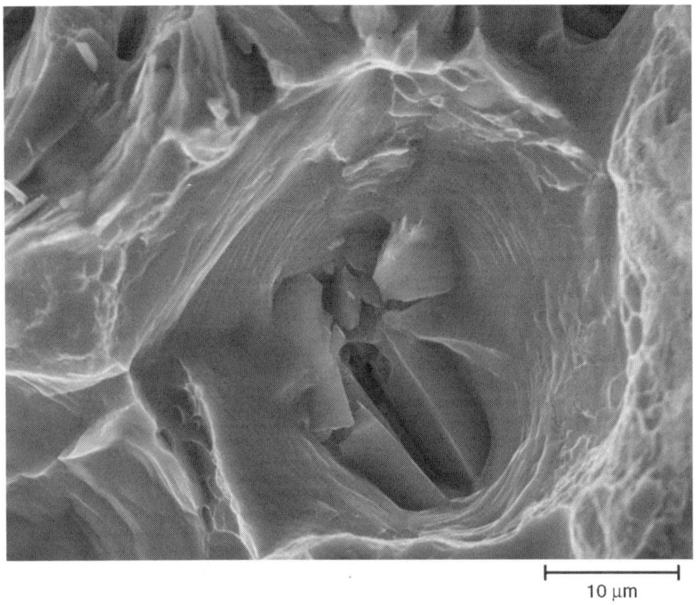

Fig. 6.66 Fracture in two-phase region. The cell is formed from a deformed matrix band around the cracked silicon particles. The zone of the interface cohesion is present on the interface between α-aluminum and silicon. In the silicon particle, the numerous cleavage cracks are visible. In the microregion of the solid solution the oval and open shear dimples are revealed. 1800×

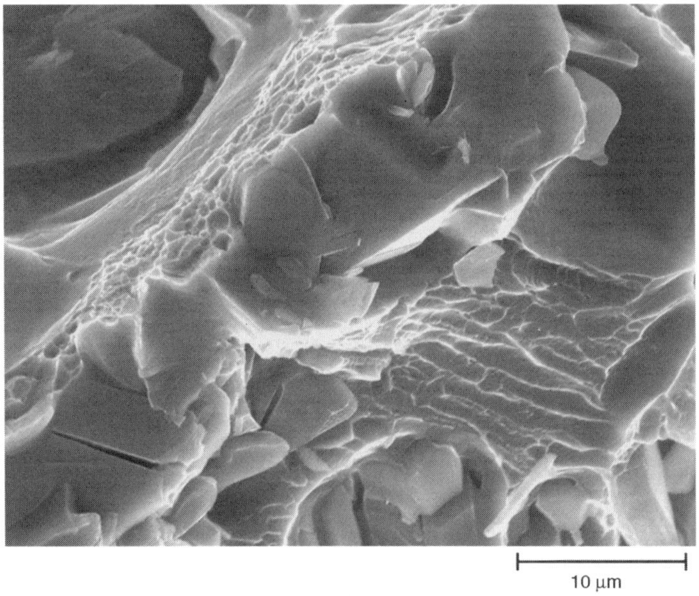

Fig. 6.67 The microregions of the fracture of mixed morphology. The cleavage facets (silicon) and shear dimples (α-aluminum) (see Fig. 6.8) are arranged in parallel bands. 2400×

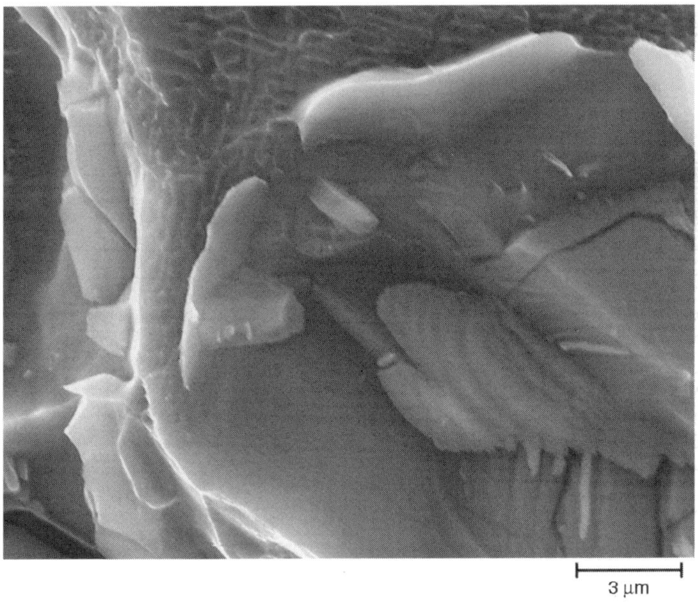

Fig. 6.68 Cleavage fracture in the silicon precipitate, characterized by the steps on the cleavage facets and the secondary cracks. 5000×

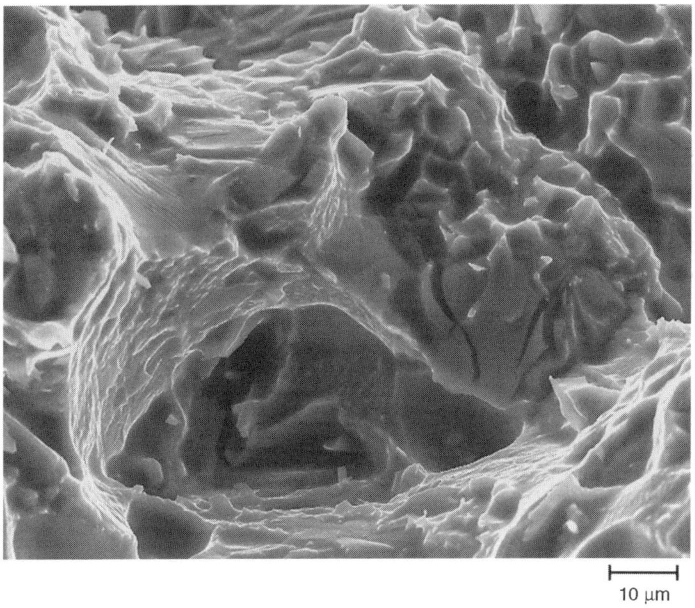

Fig. 6.69 Morphology of the fracture surface in the deformed α-aluminum solid solution. The shelves of the oval dimples are shown. 1000×

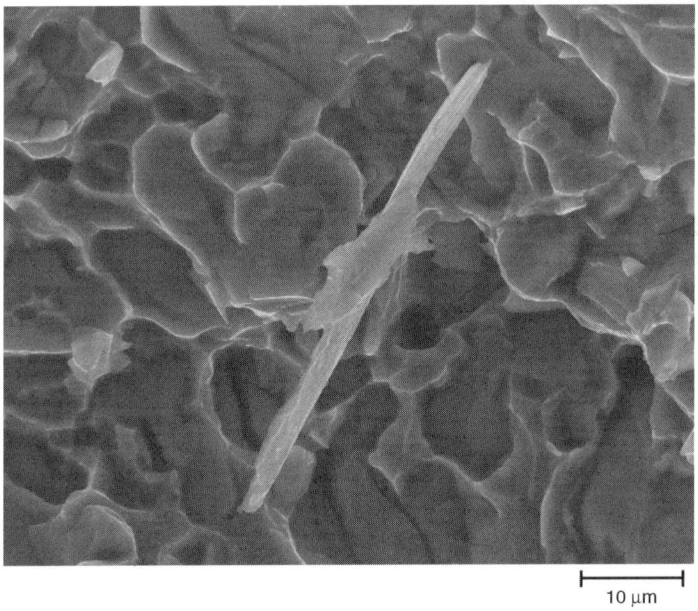

Fig. 6.70 The needle-shape precipitate of the β-AlFeSi phase on the surface of the cellular fracture can be observed. 1500×

Fracture Surfaces of Alloy 356.0 (AlSi7Mg), Refined, Modified, Permanent Mold Casting, T6, Fracture after V-Notch Impact Test, at 21 °C (70 °F)

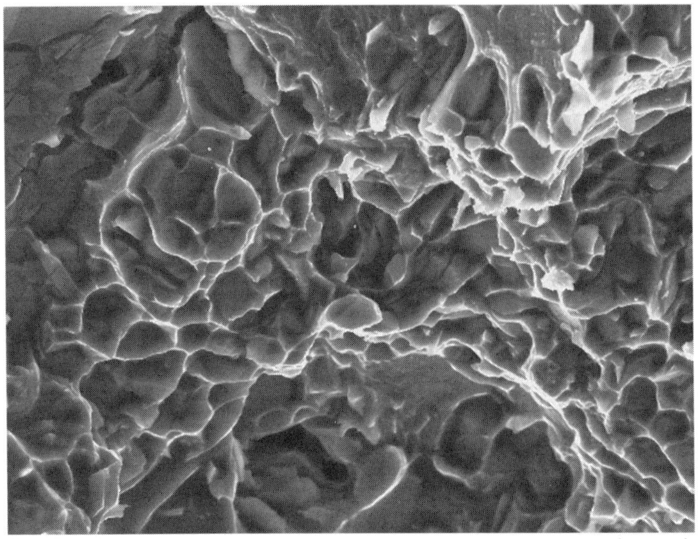

Fig. 6.71 Transcrystalline, cellular fracture in two-phase region (see Fig. 6.9). Zones of retained cohesion can be observed on the interfaces between α-aluminum and silicon. The silicon precipitates are round with micronecks of the α-aluminum solid solution. Secondary cracks in the silicon particles are shown. 1000×

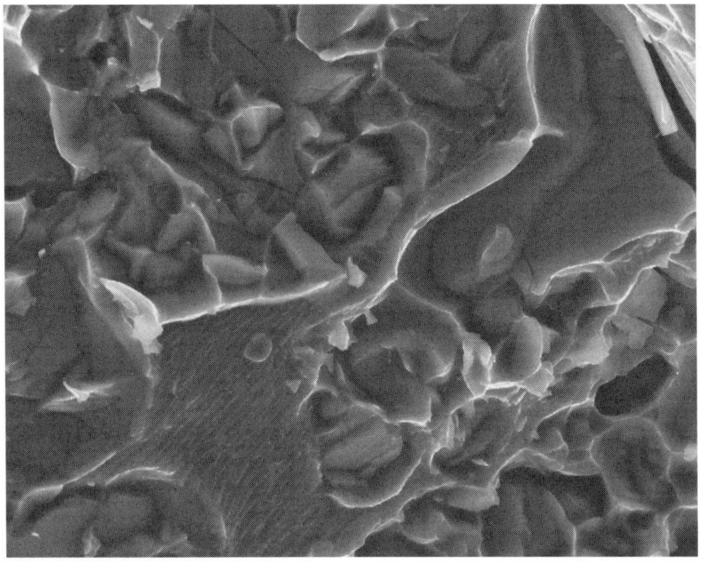

Fig. 6.72 Transcrystalline, cellular fracture in the two-phase region (see Fig. 6.11). The shear surface and cracked micronecks of the matrix among the silicon particles are visible. 1500×

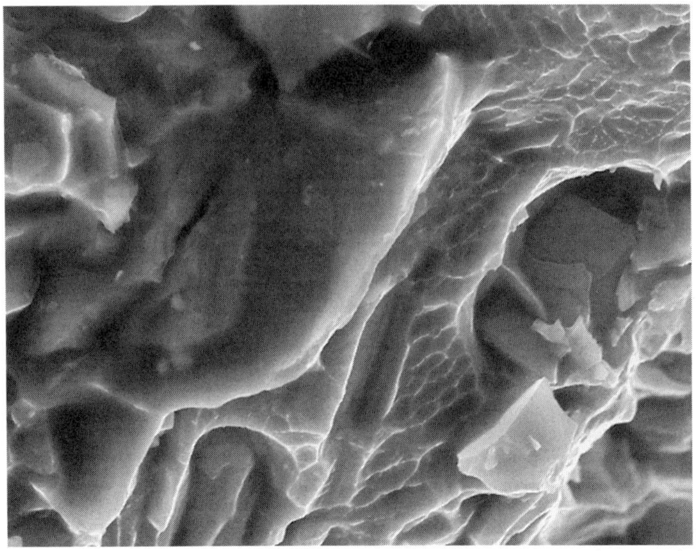

Fig. 6.73 The morphology of the shear region in α-aluminum solid solution with visible bands of the oval dimples between two silicon particles is shown. 2500×

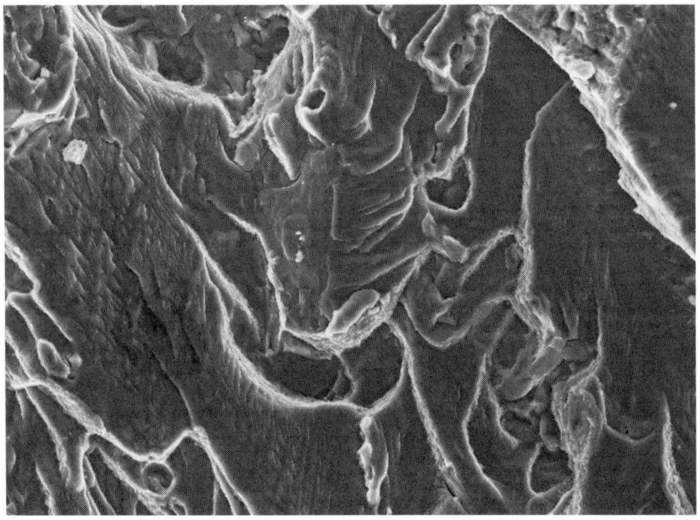

Fig. 6.74 Transcrystalline fracture. Cleavage cracks in the α-Al(FeMn)Si phase particle and in the silicon precipitate were formed. Cleavage steps on the cleavage facets can be observed. Shear decohesion took place in the matrix. 450×

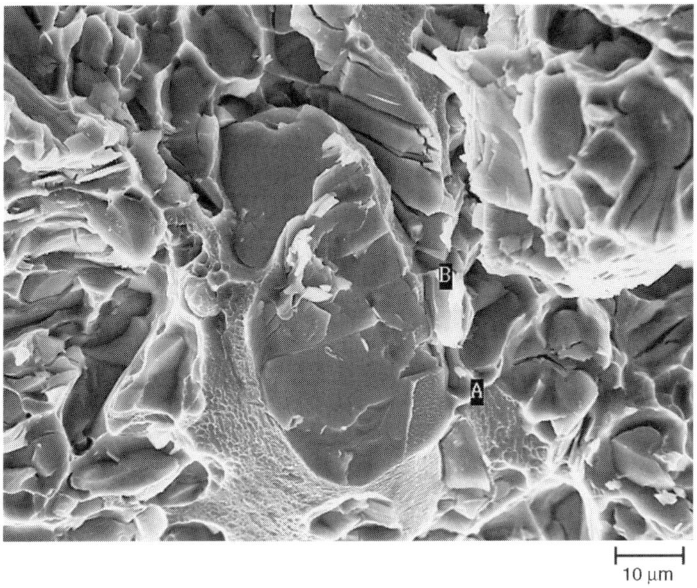

Fig. 6.75 Crack front propagated on several different cleavage planes in the silicon particle. Traces of plastic deformation are visible in the α-aluminum solid solution, around this particle. 1000×

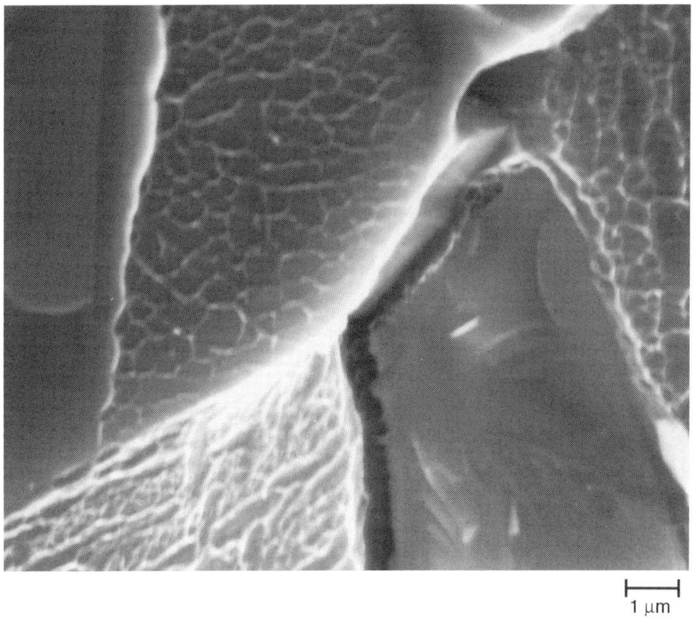

Fig. 6.76 Detail of [A] in Fig. 6.75.Fracture in two-phase region. Cleavage facets with steps in intermetallic phase $Al_8Mg_3FeSi_6$ are shown. Traces of plastic deformation in α-aluminum solid solution can be observed. The decohesion process starts on the interface between α-aluminum and silicon. 7500×

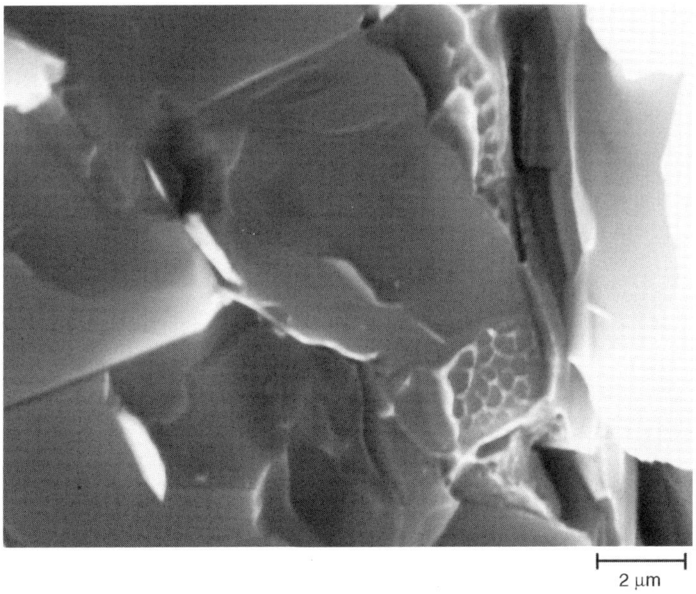

Fig. 6.77 Detail of [B] in Fig. 6.75. Cleavage steps in the silicon facets were formed. The microneck of the matrix rounds the zone of the high cohesion forces on the interface. 6500×

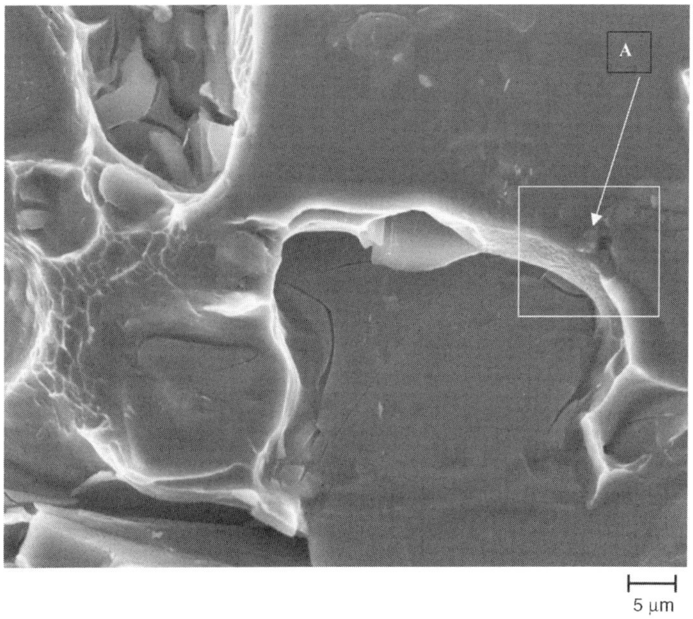

Fig. 6.78 Transcrystalline cleavage fracture developed in two neighboring silicon particles. In the α-aluminum solid solution, traces of plastic deformation are present. 1500×

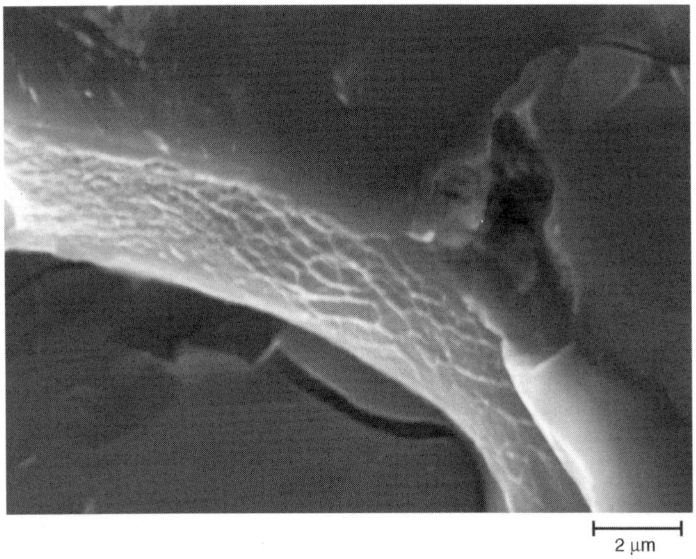

Fig. 6.79 Detail of [A] in Fig. 6.78. Steps on the cleavage facets and secondary cracks in silicon particles can be observed. In the matrix, between two silicon precipitates, the bands of the dimples resulted from plastic deformation. 6500×

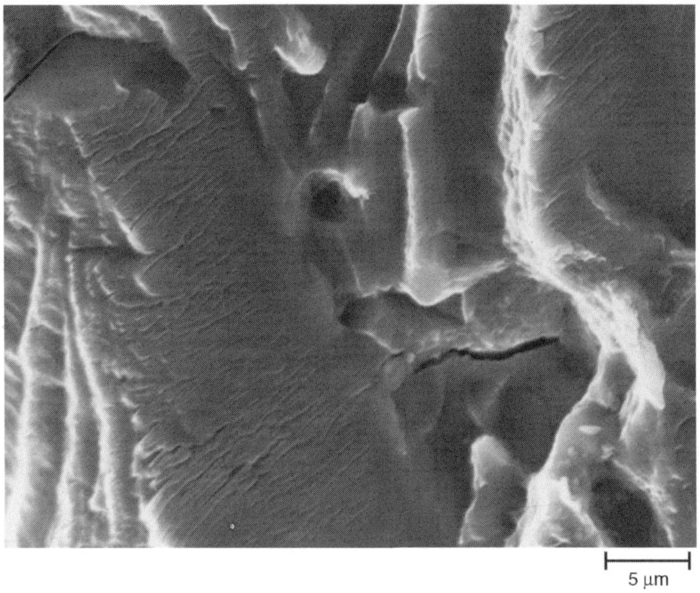

Fig. 6.80 Cracked intermetallic phase $Al_8Mg_3FeSi_6$ visible on the fracture surface. The waved steps and secondary cracks are shown. 2650×

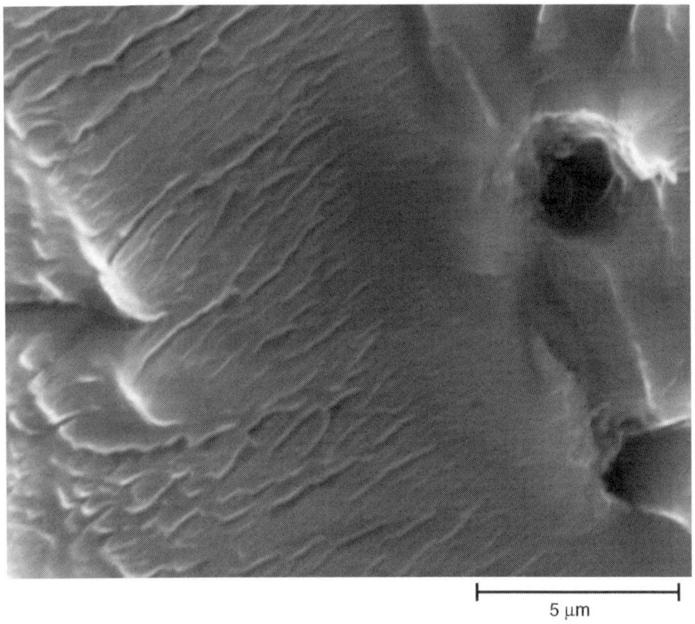

Fig. 6.81 Detail of Fig. 6.80. Steps on the cleavage facets, joining as approaching to the interface. 5400×

CHAPTER 7

Alloy 359.0 (AlSi9Mg)

Composition and Properties

Alloy	Designation	Chemical composition (main components), wt%						Properties(a)	
		Si	Cu	Mg	Mn	Ni	Fe	R_m, MPa	A_5, %
As-examined	AlSi9Mg	8.5–10.5	. . .	0.25–0.4	0.2–0.5	. . .	0.9 max	240	2.5
Aluminum Association standard(b)	359.0	8.5–9.5	0.20 max	0.0.50–0.7	0.10 max	. . .	0.20 max	310	4

(a) R_m, ultimate tensile strength; A_5, elongation measured over a length of 5.65 S_o, where S_o is the cross-sectional area of the test specimen before the test. (b) Alloy 359.0 is registered with the Aluminum Association and ASTM for sand and permanent mold casting. R_m listed is for 359.0-T61 and is the minimum ultimate tensile strength for a separately cast permanent mold specimen.

Microstructures

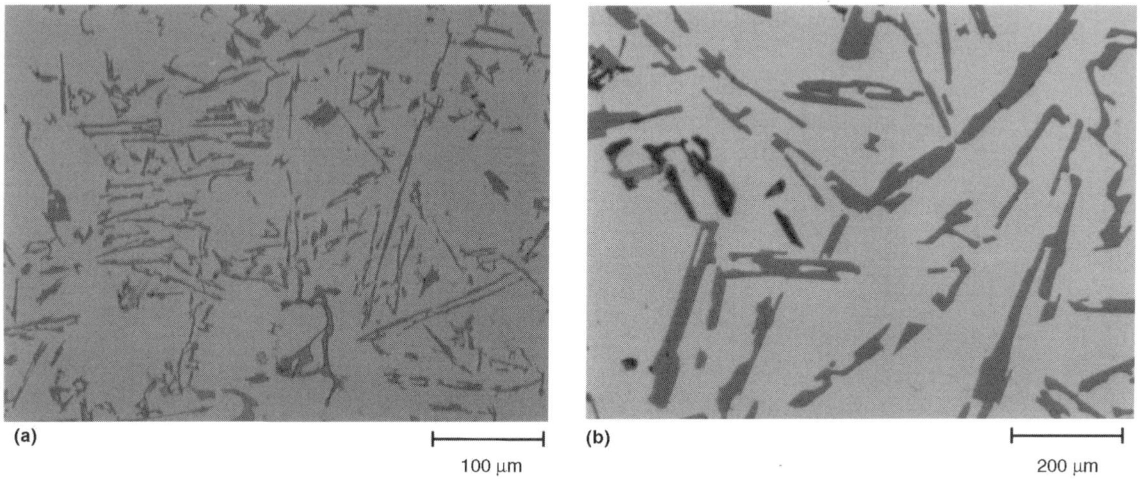

Fig. 7.1 Microstructures of AlSi9Mg (Alloy 359.0). Light microscope micrographs; etched with 1% HF. (a) As-cast (F), 150×. (b) As-cast (F), 750×

Fracture Profiles Alloy of 359.0 (AlSi9Mg), Refined, Permanent Mold Casting

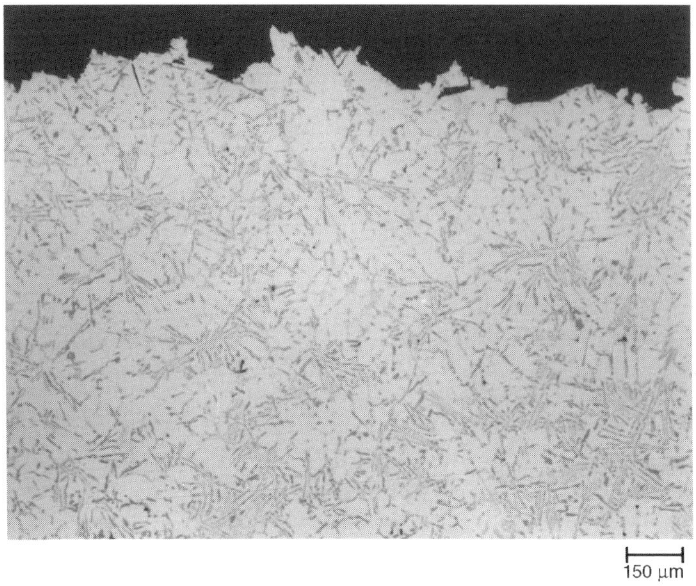

Fig. 7.2 Fracture profile of specimen after static tensile test (rim zone of the specimen). Zigzag line of the main crack reflects the morphology of the fractured micronecks, the shear edges of the α-aluminum solid solution and the cleavage lines in the precipitates of the eutectic silicon. Secondary cracks are visible as well. 50×

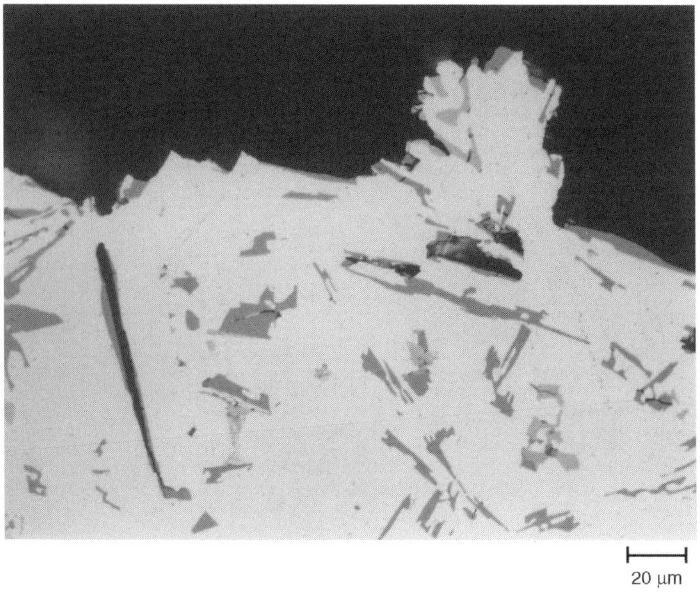

Fig. 7.3 Detail of the profile visible in Fig. 7.2. Cleavage lines in the precipitates of silicon and intermetallic phase α-Al(FeMn)Si are visible on the main profile line and in the secondary cracks. Zigzag element of the profile reveals the line of the fractured micronecks in regions of the α-aluminum solid solution. 400×

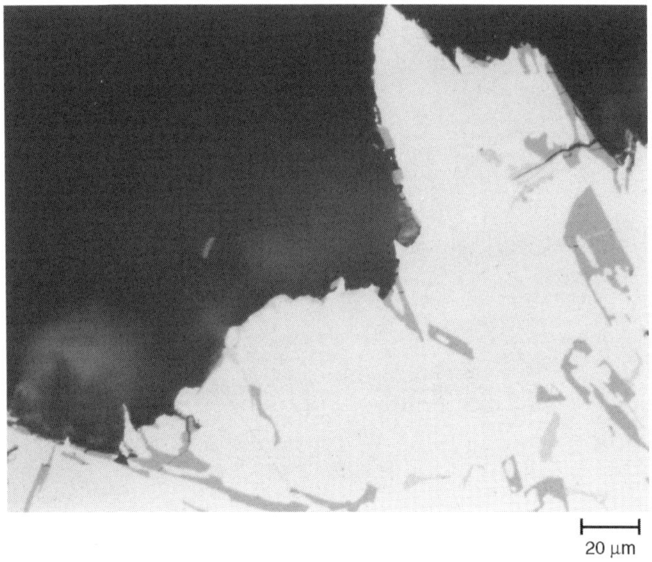

Fig. 7.4 Detail of the profile visible in Fig. 7.2. The main crack has propagated in the polyphase region. Numerous cleavage cracks are visible in silicon and the brittle intermetallic phase particles. The previously deformed micronecks were cracked in α-aluminum solid solution. 400×

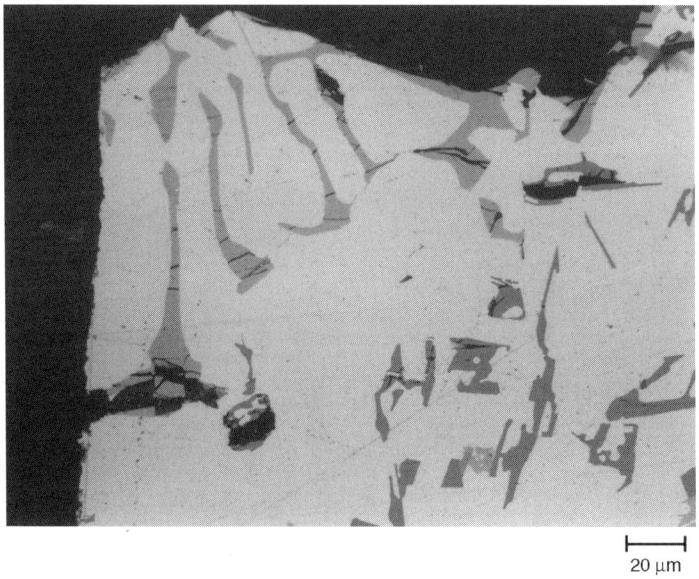

Fig. 7.5 Fracture profile of specimen after static tensile test (rim zone of the specimen) Numerous cleavage cracks are visible in the precipitates of the brittle phase α-Al(FeMn)Si on the main crack profile. The sharp microneck of α-aluminum solid solution has cracked after previous plastic deformation. On the edges of the specimen, the cleavage crack in the α-Al(FeMn)Si phase was formed. 400×

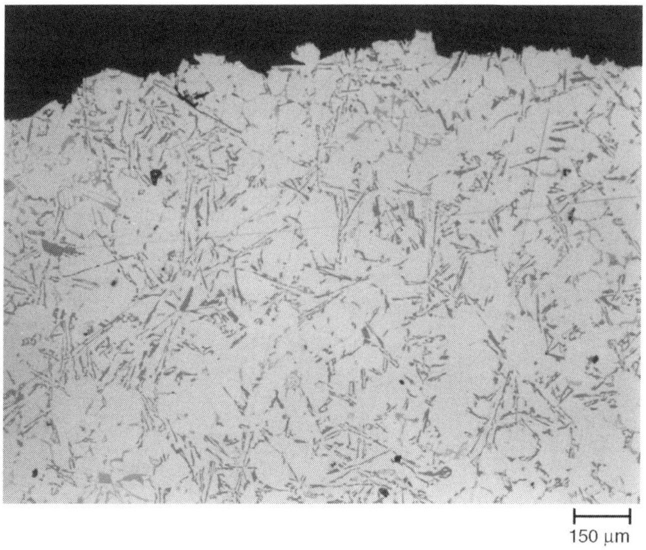

Fig. 7.6 Fracture profile of specimen after static tensile test (center zone of the specimen) in two-phase regions. 50×

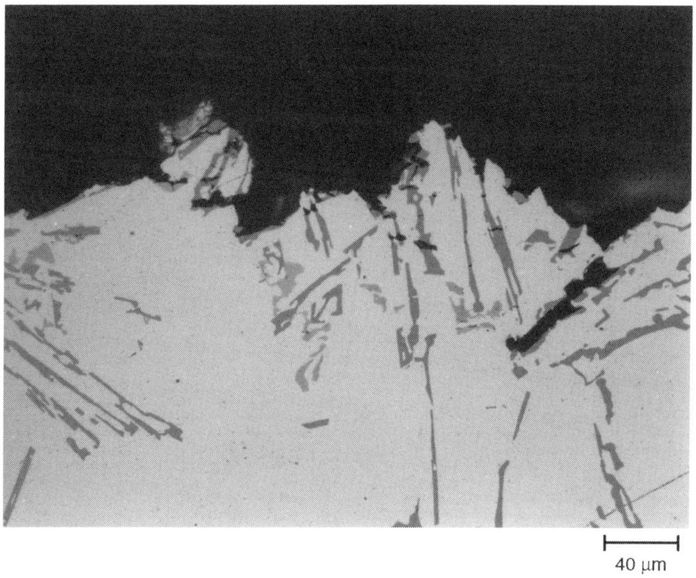

Fig. 7.7 Fracture profile of specimen after static tensile test (center zone of the specimen) in two-phase regions. The main crack front crosses polyphase, interdendritic eutectic. The secondary cracks were initiated in the brittle phase precipitates. Between brittle particles, the micronecks of the deformed α-aluminum solid solution cracked. 250×

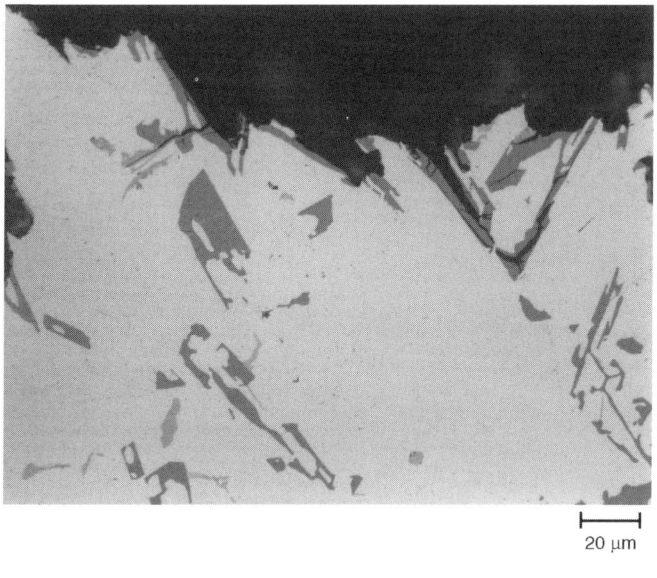

Fig. 7.8 Fracture profile of specimen after static tensile test (center zone of the specimen). The main crack crossed polyphase, interdendritic eutectic. The profile line of the main crack, of characteristic step shape, reveals the crack path on the cleavage planes of silicon particles. In the microregions of the α-aluminum solid solution, some plastic deformation of the micronecks took place. Numerous, brittle secondary cracks are visible in the brittle eutectic phases. 400×

Fig. 7.9 Fracture profile after V-notch impact test at 21 °C (70 °F). Zigzag profile of the main crack reveals the fracture path. There are visible numerous, rectilinear, secondary cracks. 50×

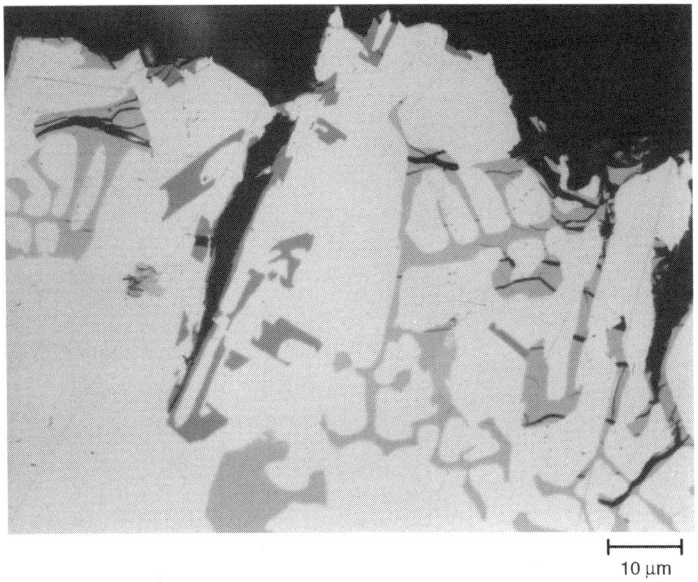

Fig. 7.10 Detail of the profile visible in Fig. 7.9. In the subsurface zone, the branched, secondary cracks are visible in the brittle eutectic phases: silicon and α-Al(FeMn)Si. On the profile line of the main crack, between regions of cleavage fracture, the sharp micronecks of the α-aluminum solid solution, previously plastically deformed, were cracked. On the shear edges, in the α-aluminum solid solution, the shear bands are visible. 1000×

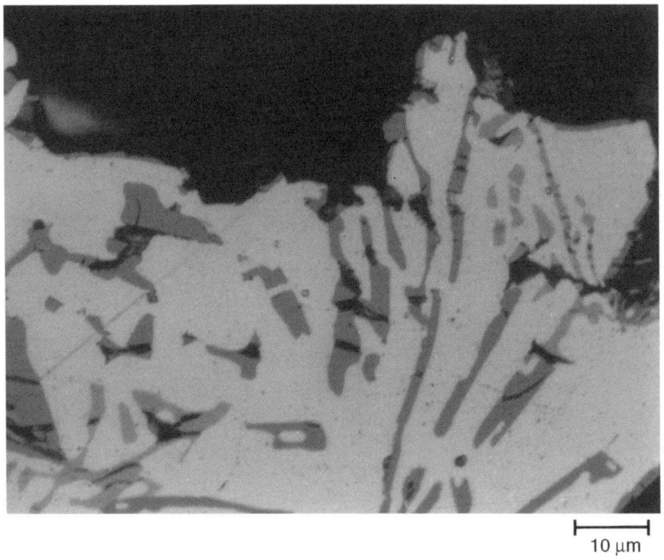

Fig. 7.11 Fracture profile after V-notch impact test at 21 °C (70 °F) in zone of the notch. The main crack crossed the cleavage planes of silicon precipitates. Numerous secondary cracks in silicon and intermetallic α-Al(FeMn)Si in fracture zone can be observed The short, sharp necks have formed in the plastically deformed α-aluminum solid solution. 1000×

Fig. 7.12 Fracture profile after V-notch impact test at 21 °C (70 °F) in rim zone of the specimen. 50×

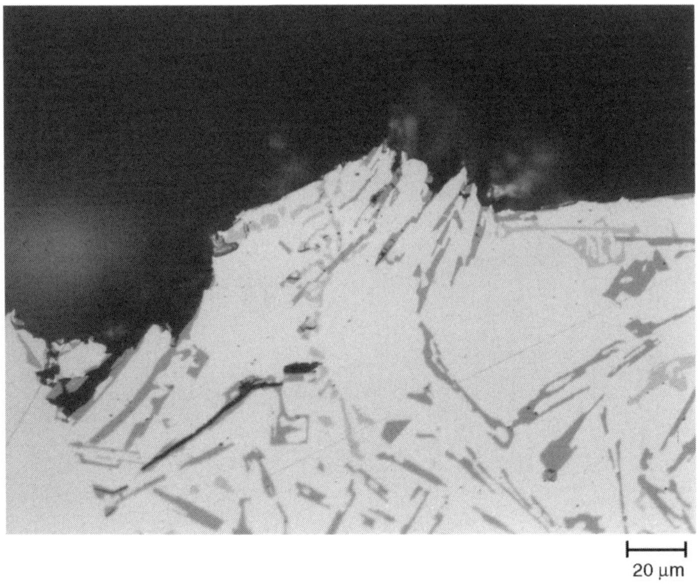

Fig. 7.13 Detail of the profile visible in Fig. 7.12. On the profile line of the main crack the cleavage lines are visible (long, rectilinear parts of the crack path on the cleavage planes of the brittle constituents of the interdendritic eutectic). The micronecks of the locally deformed α-aluminum solid solution are situated between brittle particles. 400×

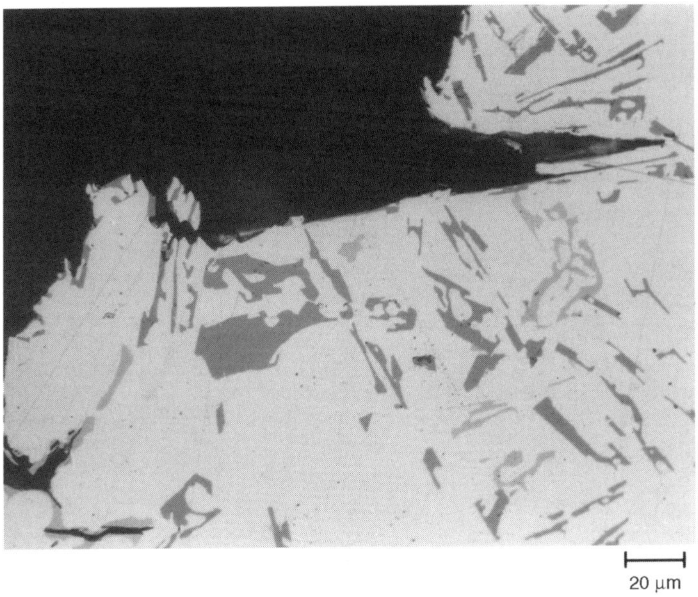

Fig. 7.14 Detail of the profile from Fig. 7.12. Screen shades the long part of the main crack profile, formed by the cleavage line—a trace of the crack path in the cleavage plane of the silicon precipitate. 400×

Fig. 7.15 Fracture profile after V-notch impact test at 21 °C (70 °F) in the central zone of the specimen. 50×

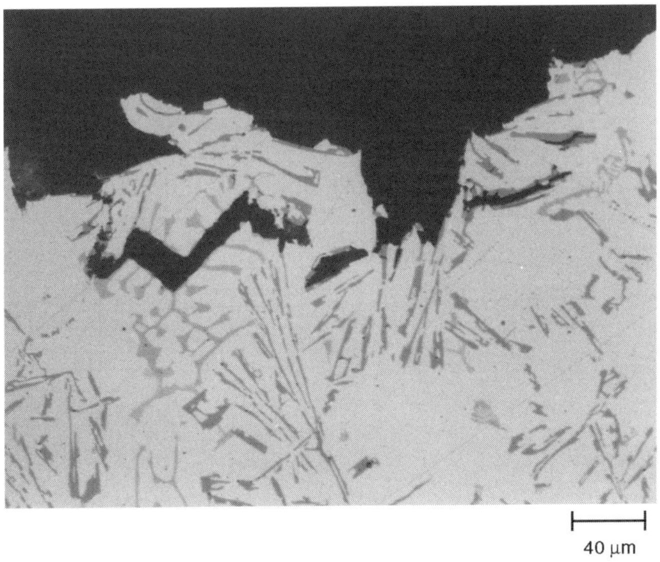

Fig. 7.16 Detail of the profile from Fig. 7.15.The characteristic zigzag crack is visible in the two-phase region α-aluminum + α-Al(FeMn)Si. The main crack front crosses the cleavage planes of the eutectic silicon particles, among which the sharp micronecks of the deformed α-aluminum solid solution are visible. 250×

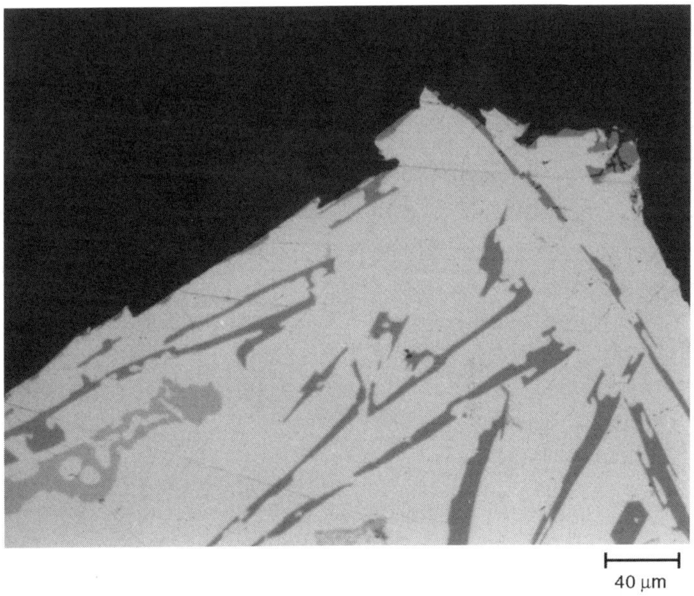

Fig. 7.17 Fracture profile of specimen after V-notch impact test at 21 °C (70 °F) in central zone of specimen. The main crack crossed the cleavage planes of the silicon particles. The steps caused by the by-pass of the crack front into the parallel cleavage planes in the neighboring particle (left part of the micrograph). In the α-aluminum solid solution regions the fractured micronecks are visible. 250×

Fracture Surfaces of Alloy 359.0 (AlSi9Mg), Refined, Modified, Permanent Mold Casting, Fracture after Static Tensile Test

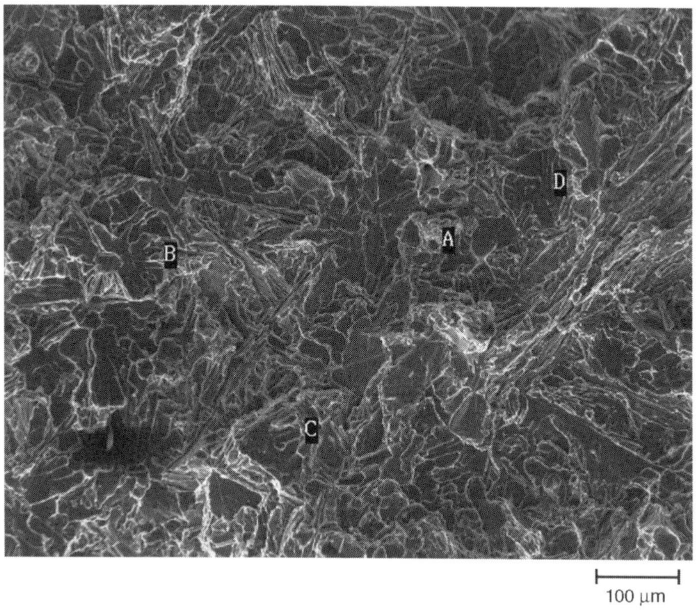

Fig. 7.18 Transcrystalline fracture of greatly developed surface. Fracture areas of cleavage and cellular morphology are visible (see Fig. 7.2). 120×

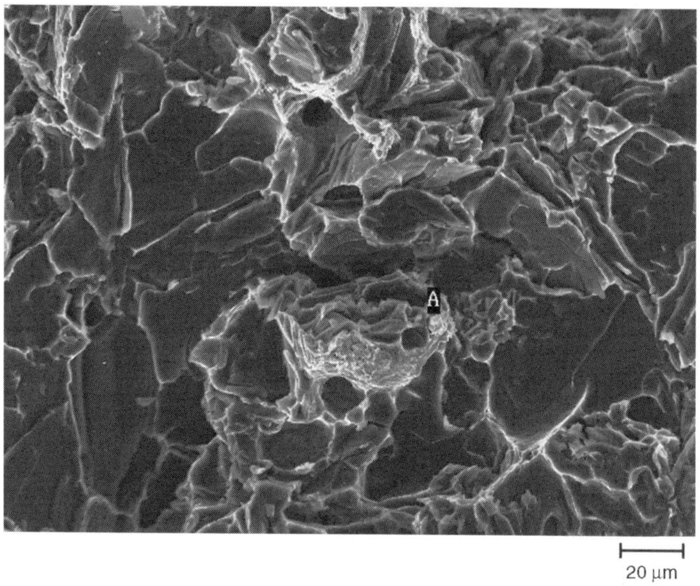

Fig. 7.19 Detail of [A] in Fig. 7.18. Transcrystalline fracture in the brittle constituents of the eutectic α-aluminum + silicon is shown. On the cleavage facets, both the rivers and the cleavage steps are present. In the matrix, the dimples and the micronecks, deformed before crack, can be observed. 450×

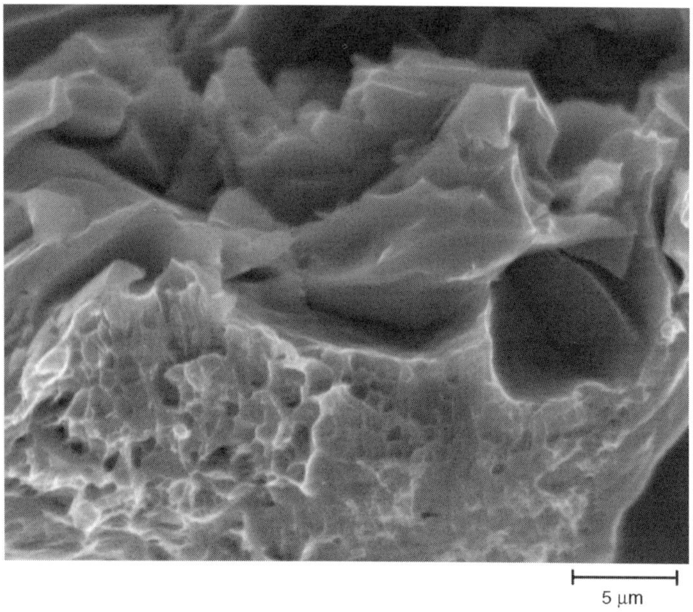

Fig. 7.20 Detail of [A] in Fig. 7.19. Oval and equiaxial dimples are situated in the deformed region of the matrix. 3000×

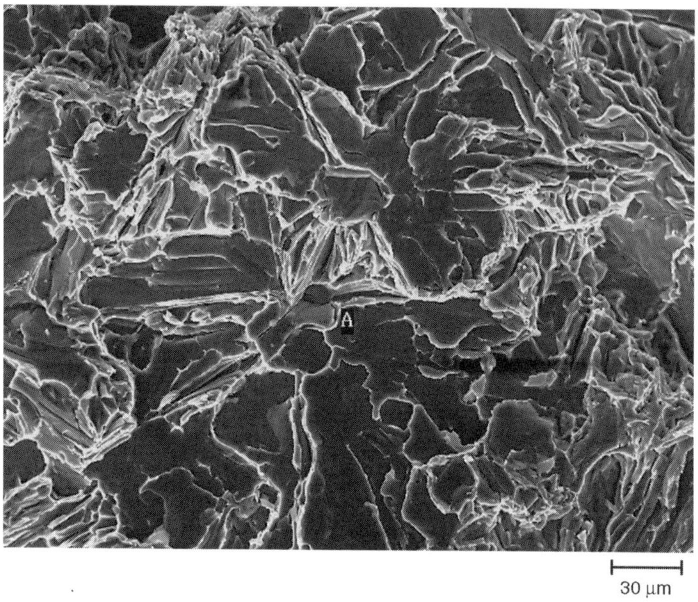

Fig. 7.21 Detail of [B] in Fig. 7.18. The crack crossed the cleavage planes in the eutectic silicon particles, parallel to the average fracture plane in this microregion (see Fig. 7.8). Cleavage steps, rivers, and tongues also are visible. 250×

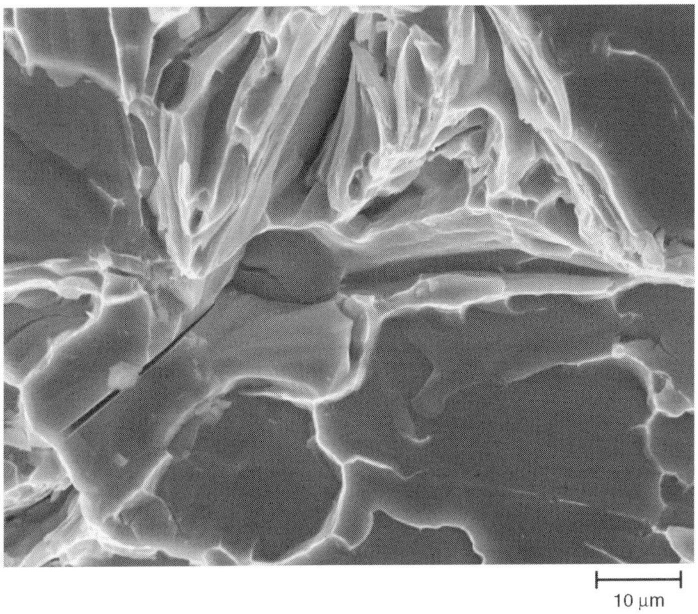

Fig. 7.22 Detail of [A] in Fig. 7.21. Cleavage facets in silicon particle are parallel to the average fracture plane in this microregion (see Fig. 7.8). Rivers, weakly developed river patterns, and secondary cracks also are shown. 1200×

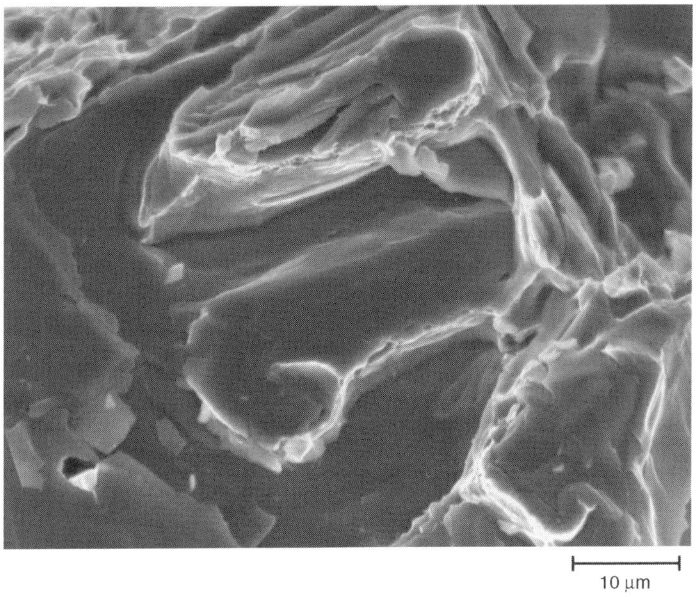

Fig. 7.23 Detail of [C] in Fig. 7.18. Transcrystalline fracture in a silicon precipitate. The crack front crosses the parallel cleavage planes. In the microregions of the α-aluminum solid solution the dimples, as a result of plastic deformation, were formed. 1500×

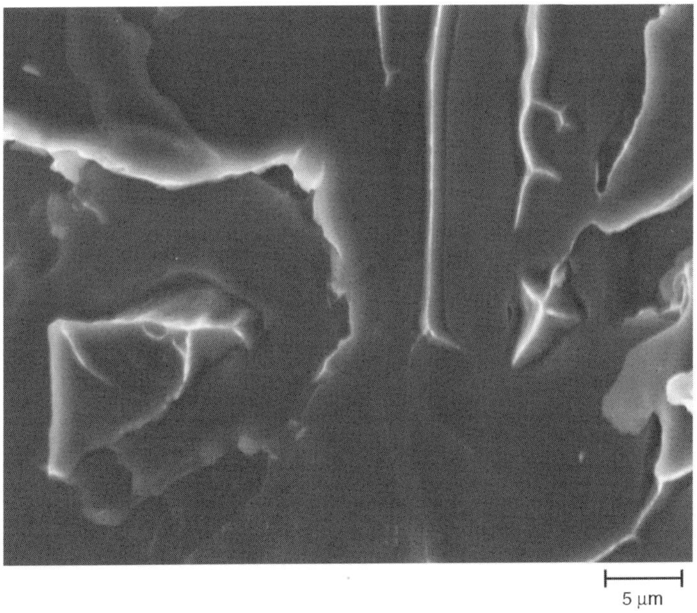

Fig. 7.24 Detail of [D] in Fig. 7.18. The branching of the cleavage steps has formed on the cleavage facet in the silicon precipitate. 2200×

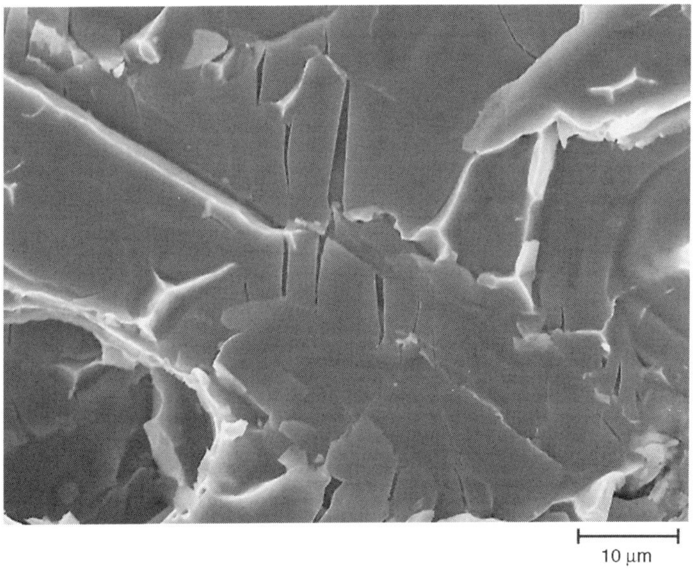

Fig. 7.25 The branching of the cleavage steps on the cleavage facets is a result of the crack propagation on several of the cleavage planes of the intermetallic phase α-aluminum (FeMn)Si precipitate. Secondary cracks also are shown in this particle (see Fig. 7.5). 1500×

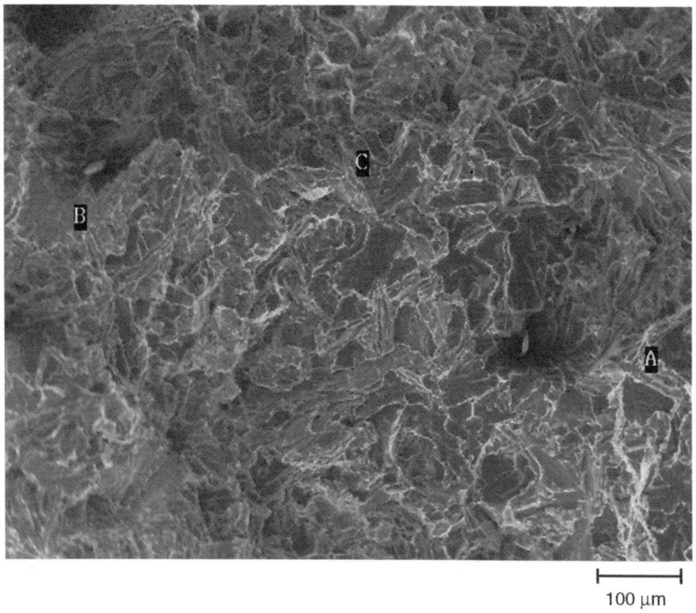

Fig. 7.26 Transcrystalline fracture of mixed morphology and greatly developed surface. Numerous cleavage cracks are visible in brittle constituents of the eutectic. The areas with cellular morphology also can be observed (see Fig. 7.2). Details of regions [A], [B], and [C] are found in Fig. 7.27, 7.28, and 7.29, respectively. 125×

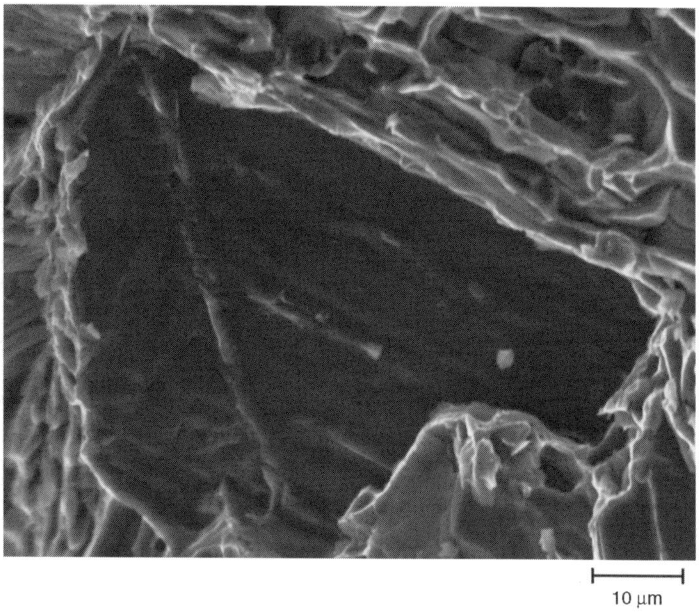

Fig. 7.27 Detail of [A] in Fig. 7.26. Transcrystalline cleavage fracture. The crack propagated on two cleavage planes in the α-Al(FeMn)Si phase precipitate. 1300×

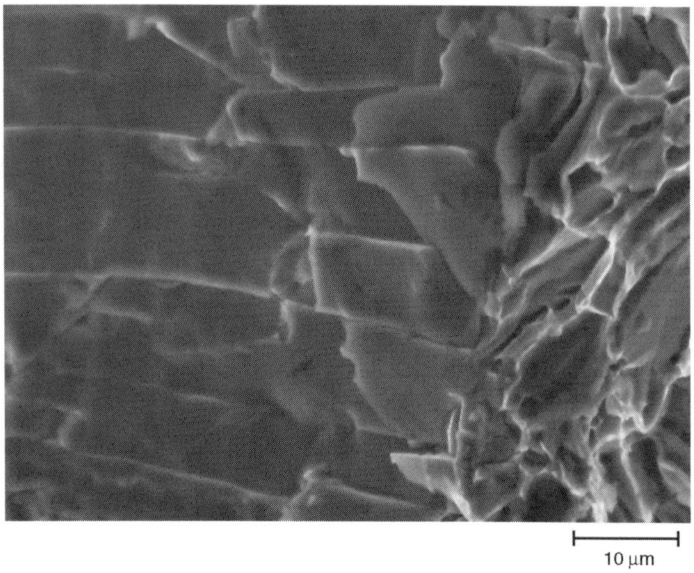

Fig. 7.28 Detail of [B] in Fig. 7.26. The surface of the cracked silicon precipitate is shown; the bands of the parallel steps on the cleavage facet are visible. In the right part of the micrograph, the elongated dimples and cells in the matrix can be observed. 1500×

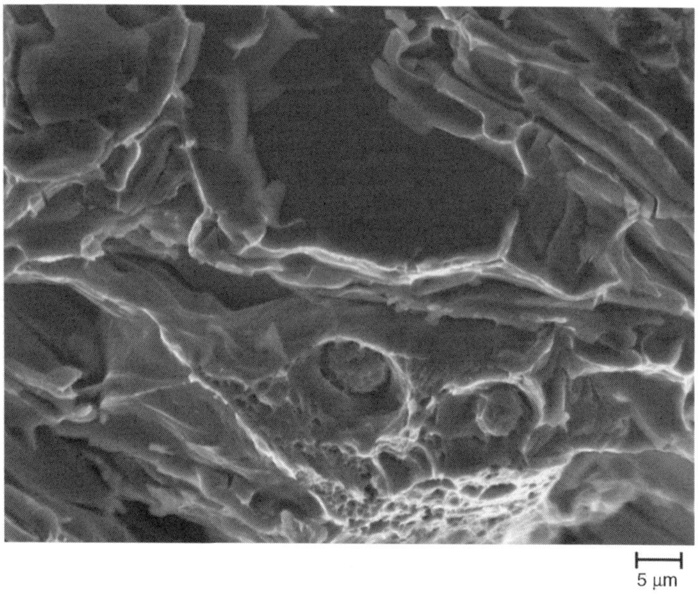

Fig. 7.29 Detail of [C] in Fig. 7.26. The step bands between parallel cleavage facets formed in the silicon particles. In the shear region of the matrix, the open and oval dimples are present. 1300×

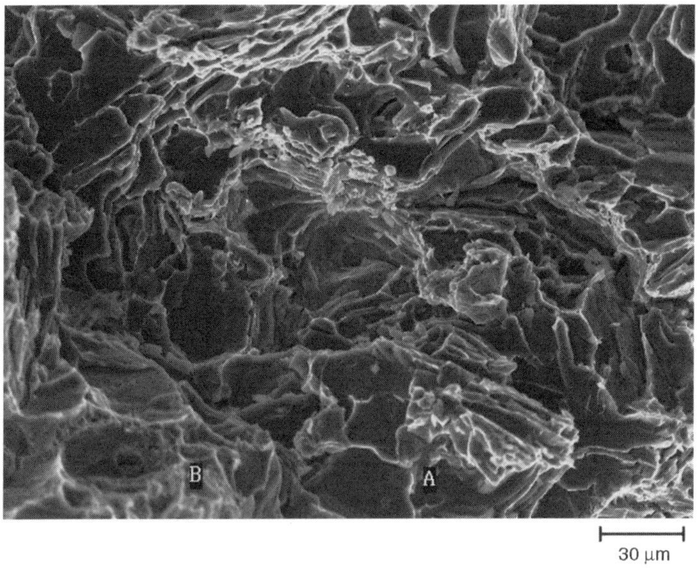

Fig. 7.30 Transcrystalline fracture of different morphology and medium-developed surface. 400×

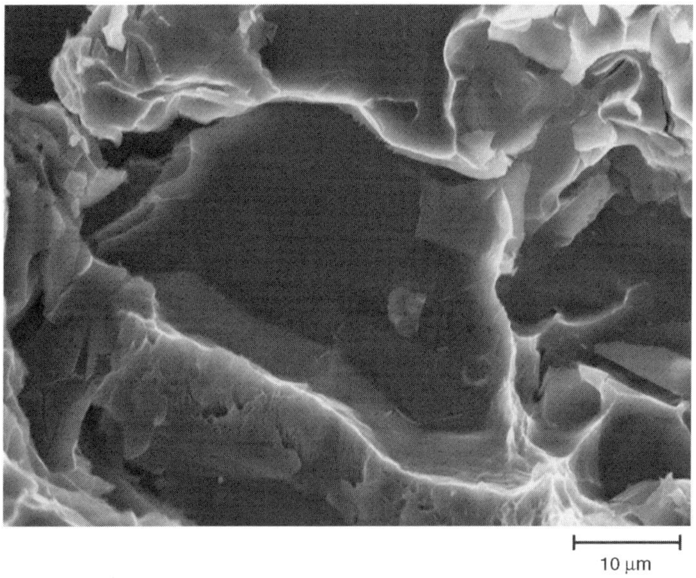

Fig. 7.31 Detail of [A] in Fig. 7.30. Cleavage crack propagated on cleavage planes in silicon particles. The tear ridge in the matrix, between two silicon particles, represents composed fracture mechanism in this microregion. The open shear dimples can be observed on the shear surface. 1500×

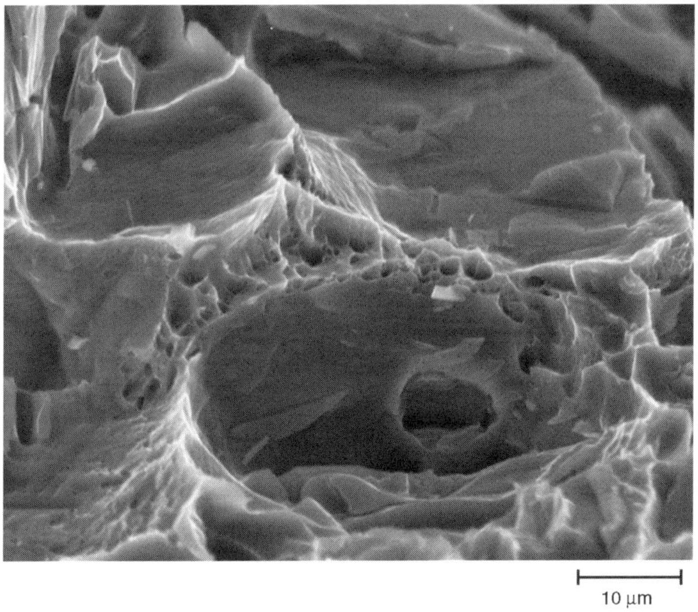

Fig. 7.32 Detail of [B] in Fig. 7.30. The area of the deformed matrix is visible around silicon particles, cracked in the cleavage planes. The bands of the equiaxial dimples were formed in the deformed α-aluminum solid solution zone. Micronecks cracked after plastic deformation are situated in the shear area. 1500×

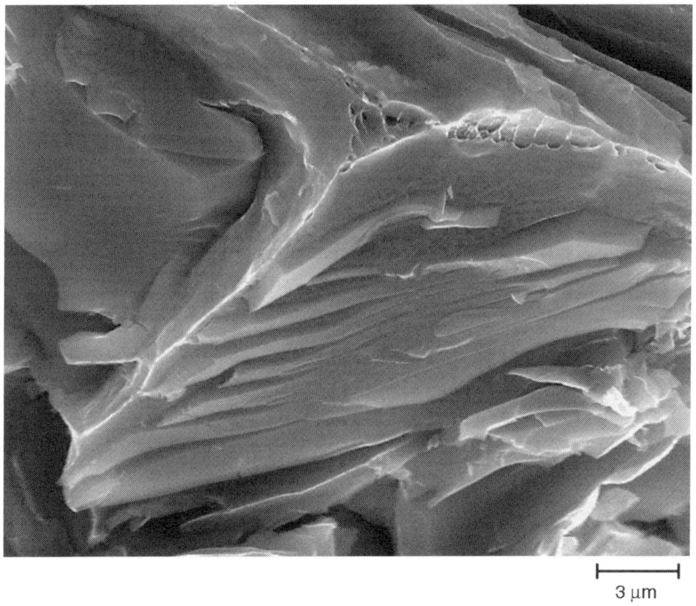

Fig. 7.33 Open dimples in the crack zone of the microneck are an effect of the ductile fracture of the matrix. 4000×

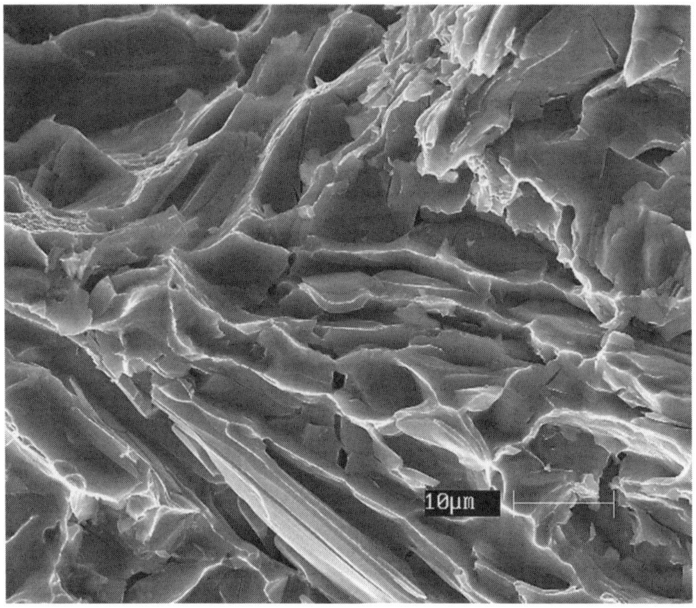

Fig. 7.34 Transcrystalline fracture of cellular morphology. The crack front crosses the zone of the interdendritic eutectic. In the matrix, on the tear ridges of α-aluminum solid solution, the dimples can be observed. Secondary cracks are present in the brittle, eutectic phases. 1500×

Fig. 7.35 Transcrystalline fracture of cellular morphology. The cells have formed the characteristic rosette, where the crack initiation took place. The interface cohesion on the interface between α-aluminum and silicon was retained. 3200×

Alloy 359.0 (AlSi9Mg), Refined, Modified, Permanent Mold Casting, Fracture after V-Notch Impact Test at 21 °C (70 °F)

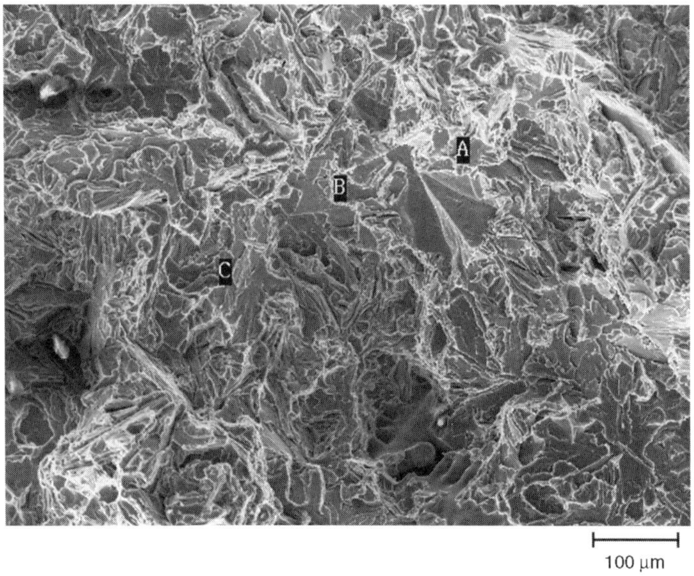

Fig. 7.36 Transcrystalline fracture of medium-developed surface. The main crack propagated by the cleavage planes in the brittle eutectic phases. The area of the cellular fracture in two-phase region also is visible. 125×

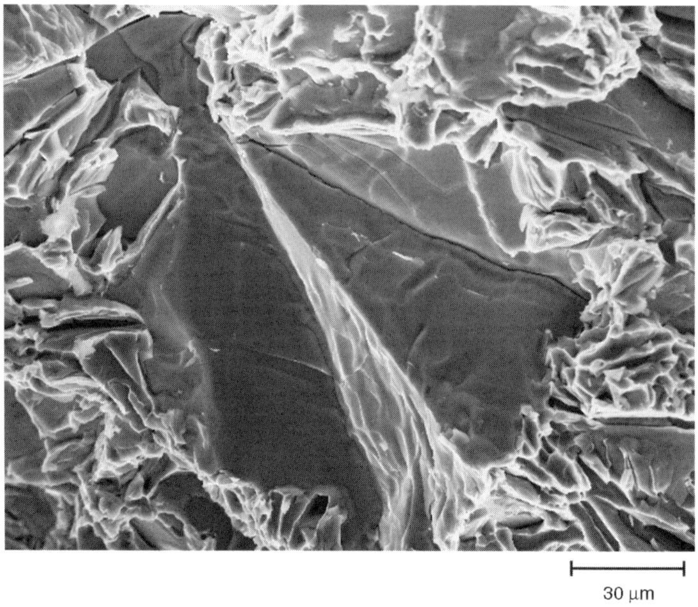

Fig. 7.37 Detail of [A] in Fig. 7.36. The crack crossed three cleavage planes in the eutectic silicon particle. Cleavage step bands also are visible. 500×

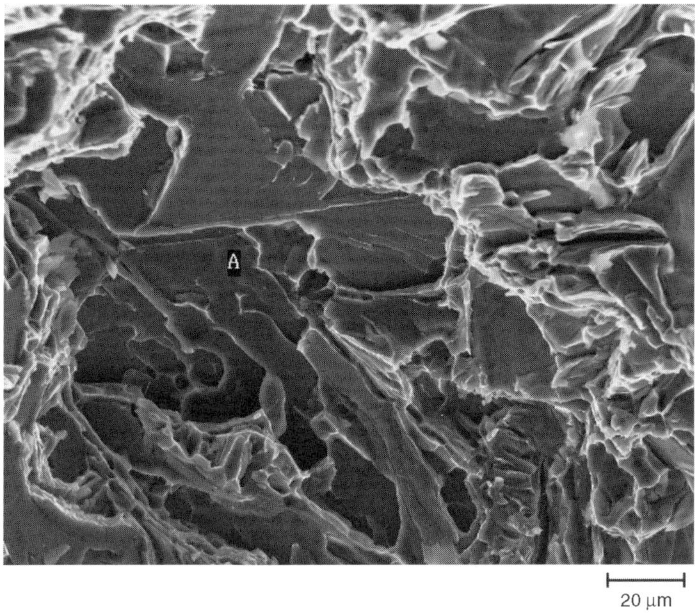

Fig. 7.38 Detail of [B] in Fig. 7.36. The crack formed on the cleavage planes in branched precipitate of the brittle α-Al(FeMn)Si phase. Cleavage steps form bands and river patterns. 550×

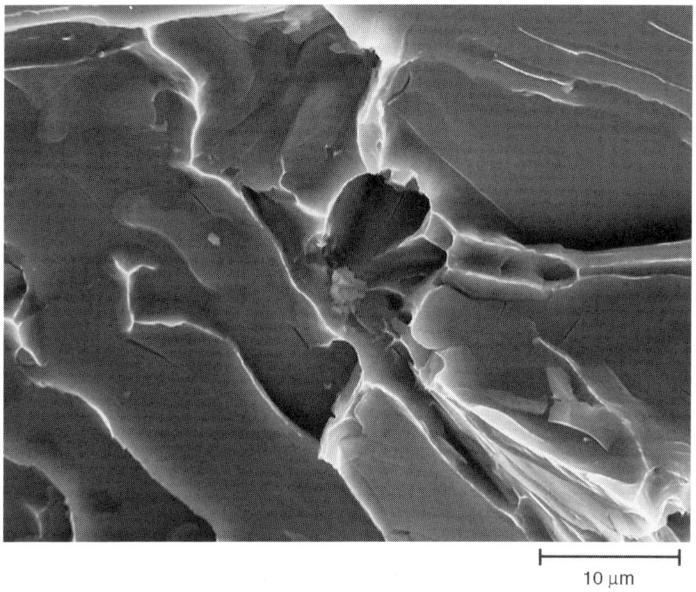

Fig. 7.39 Detail of [A] in Fig. 7.38. Cleavage steps on the cleavage facets in α-Al(FeMn)Si phase and secondary cracks on the interface can be observed. 1800×

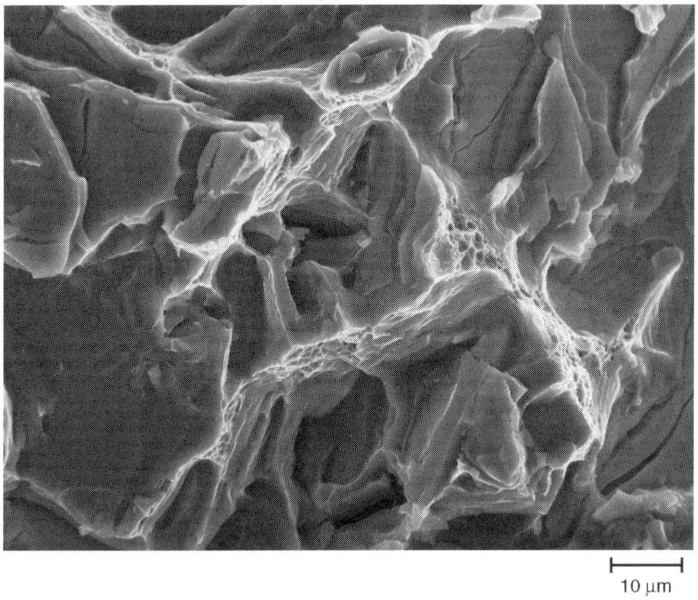

Fig. 7.40 Detail of [C] in Fig. 7.36. In the microregions of the matrix, among the silicon precipitates the shear process took place. The micronecks and oval dimples, characteristic for shear process, are visible. 1000×

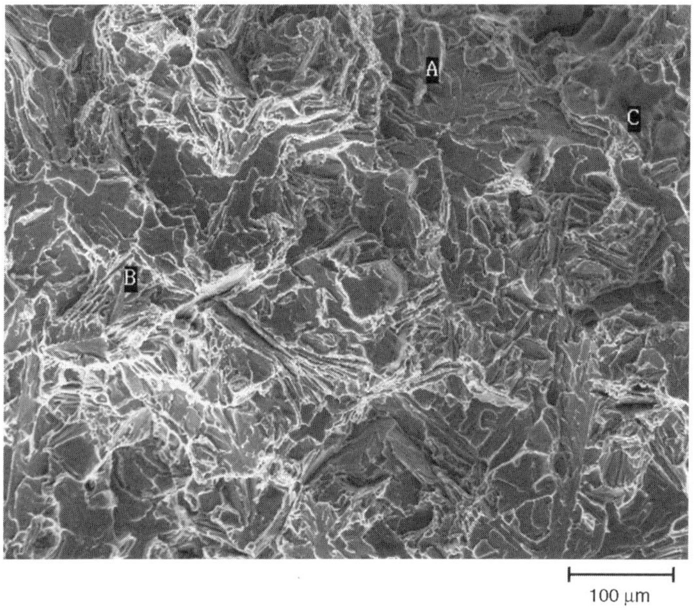

Fig. 7.41 Transcrystalline fracture of medium-developed surface. The main crack propagated on the cleavage planes in the brittle eutectic phases: silicon and intermetallic phases. 150×

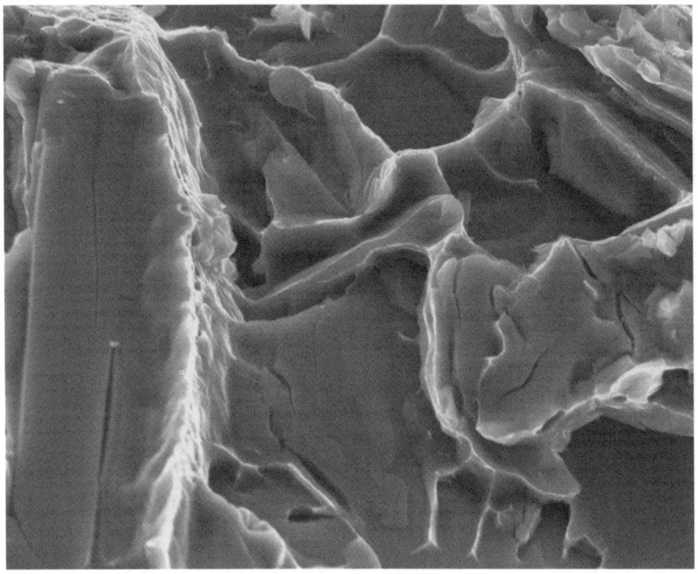

Fig. 7.42 Detail of [A] in Fig. 7.41. Transcrystalline, brittle fracture. The terraces of parallel cleavage facets are visible in the silicon precipitates, separated with the cleavage steps and secondary cracks. In the left part of the micrograph, the particle of the intermetallic β-AlFeSi phase is visible. 1350×

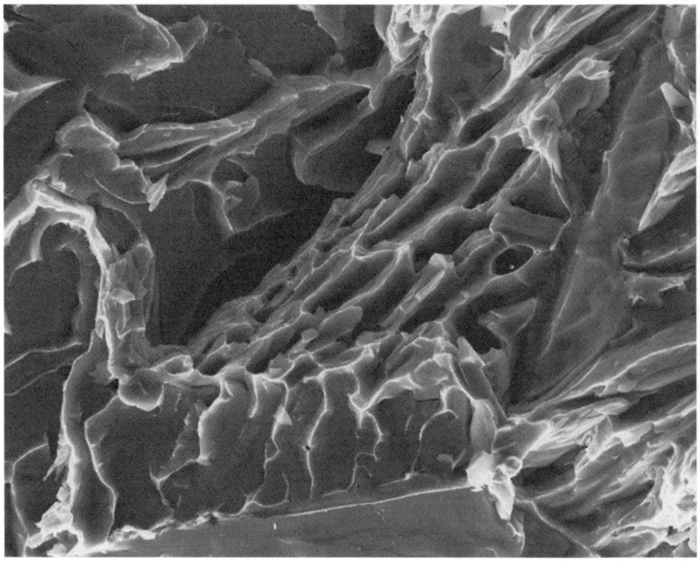

Fig. 7.43 Detail of [B] in Fig. 7.41.The crack propagated in the polyphase eutectic α-Al + Si + Mg_2Si + α-Al(FeMn)Si. The cleavage facets and the secondary cracks are separated with the steps. 1000×

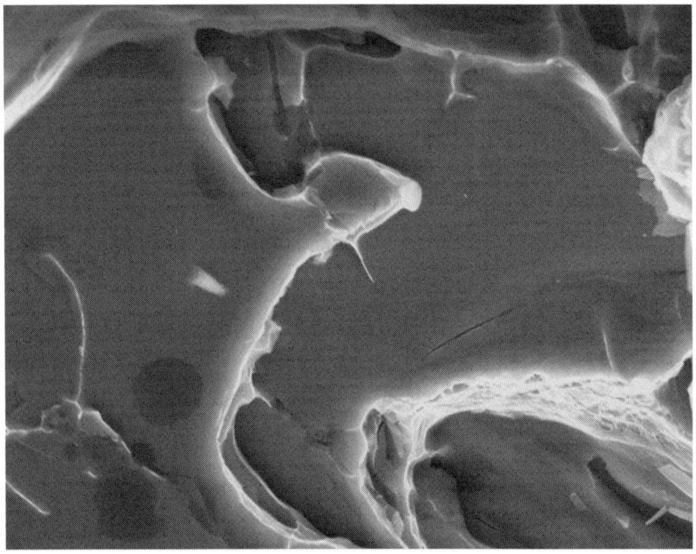

Fig. 7.44 Detail of [C] in Fig. 7.41. Transcrystalline cleavage fracture, in silicon precipitate. The steps have formed among three parallel cleavage planes and secondary cracks (see Fig. 7.17). 2000×

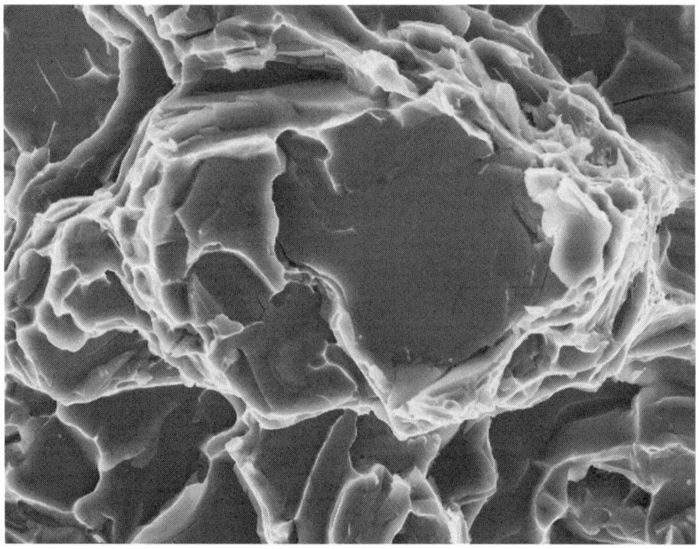

Fig. 7.45 Transcrystalline fracture. The parallel cleavage facets in the silicon precipitates are separated with either steps of different height or secondary cracks. The very small dimples observed in the matrix are an effect of plastic deformation. 1000×

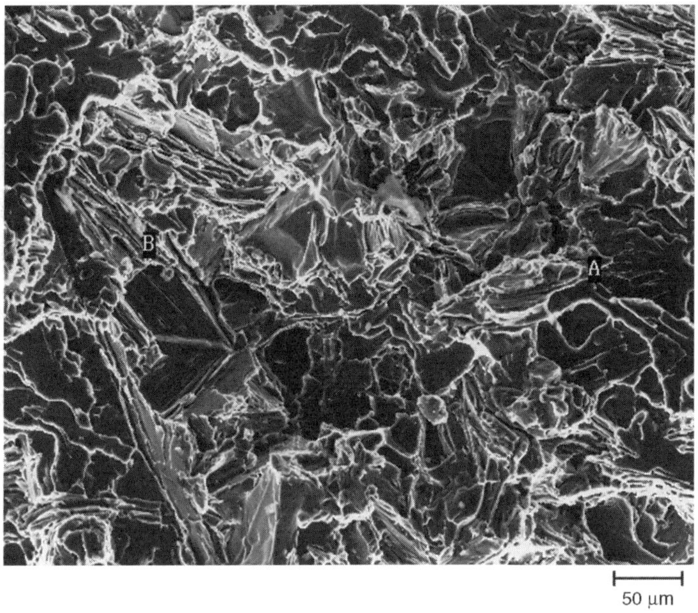

Fig. 7.46 Transcrystalline fracture of a medium-developed surface. The crack front crosses the cleavage planes in the eutectic silicon precipitates. The areas of fracture with cellular morphology also are visible. 200×

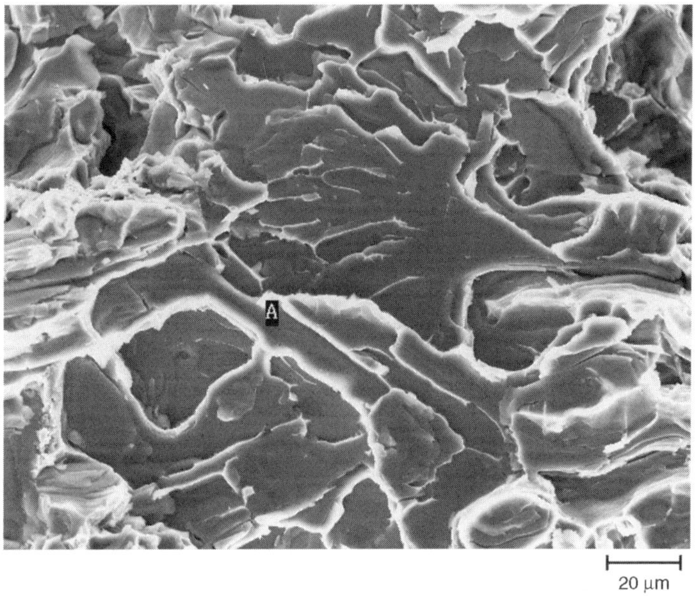

Fig. 7.47 Detail of [A] in Fig. 7.46. The crack front propagated in the silicon precipitates on the parallel cleavage planes, separated with the cleavage steps. 500×

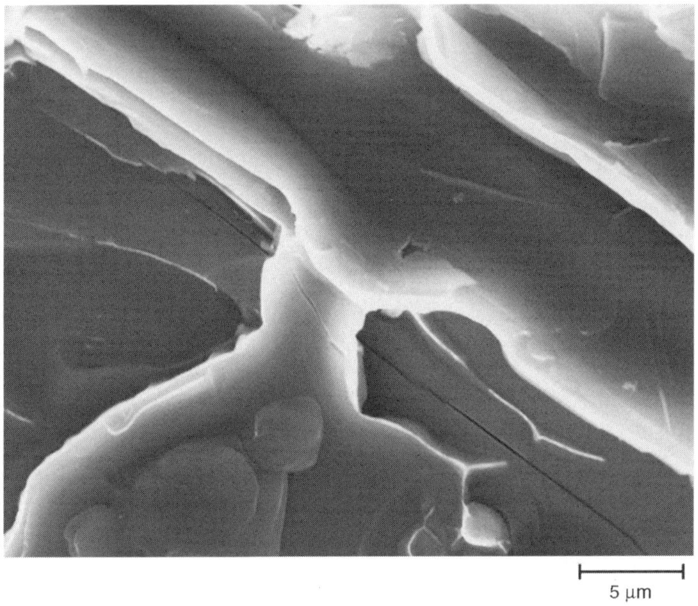

Fig. 7.48 Detail of [A] in Fig. 7.47. The steps morphology in the zone of the crack front is characteristic for the propagation on the successive cleavage planes. 3000×

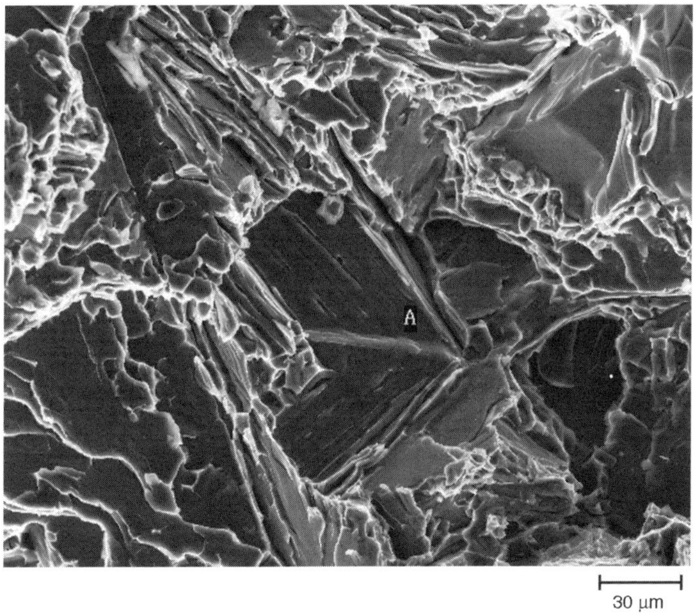

Fig. 7.49 Detail of [B] in Fig. 7.46. Transcrystalline cleavage fracture. Cracks have formed on the cleavage planes in the brittle eutectic constituents silicon and α-Al(FeMn)Si. 400×

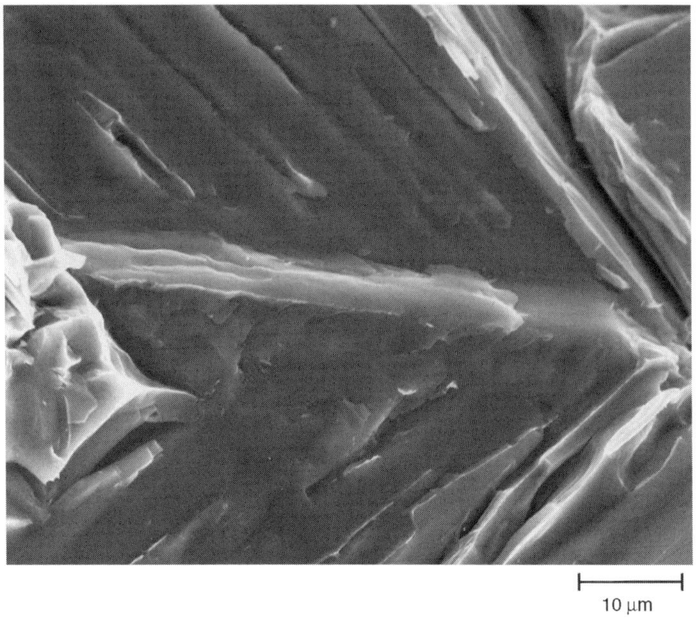

Fig. 7.50 Detail of [A] in Fig. 7.49. The parallel steps form the bands on the cleavage facets in the silicon particle. The interface cohesion was interrupted in some zones on the interface between α-aluminum and silicon. 1500×

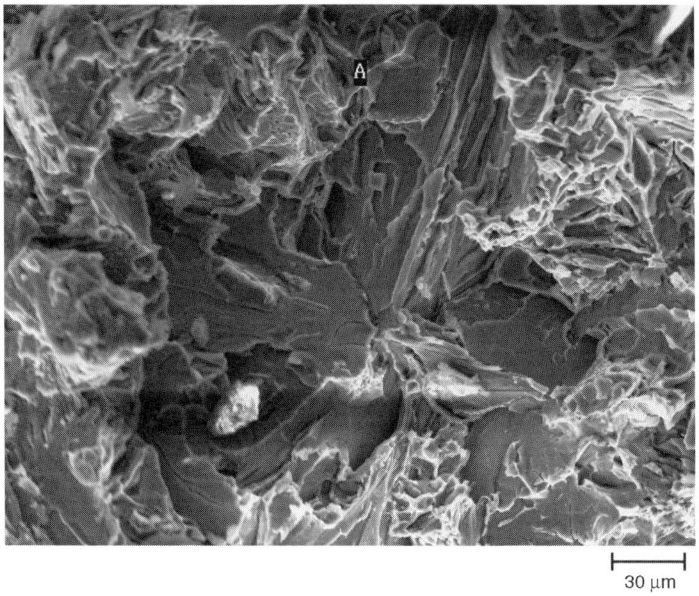

Fig. 7.51 Transcrystalline fracture of medium-developed surface. The crack crossed the cleavage planes in the brittle eutectic phases and the zones of the deformation in the matrix. The rosette of the cleavage steps on the cleavage facets forms the center of the crack initiation zone. 300×

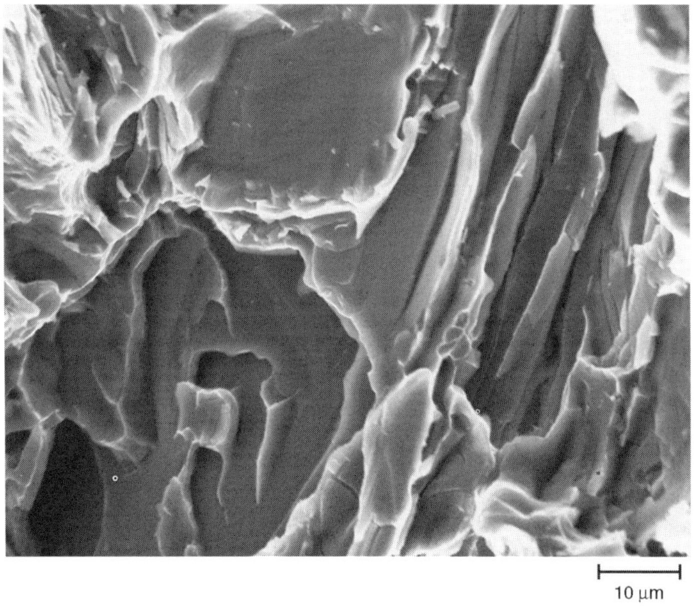

Fig. 7.52 Detail of [A] in Fig. 7.51.Cleavage steps among the parallel cleavage facets have formed in the silicon particle. In the top of the micrograph the brittle crack in the intermetallic β-AlFeSi phase (in platelike shape) is visible. The dimples were formed in the matrix as a result of plastic deformation. 1200×

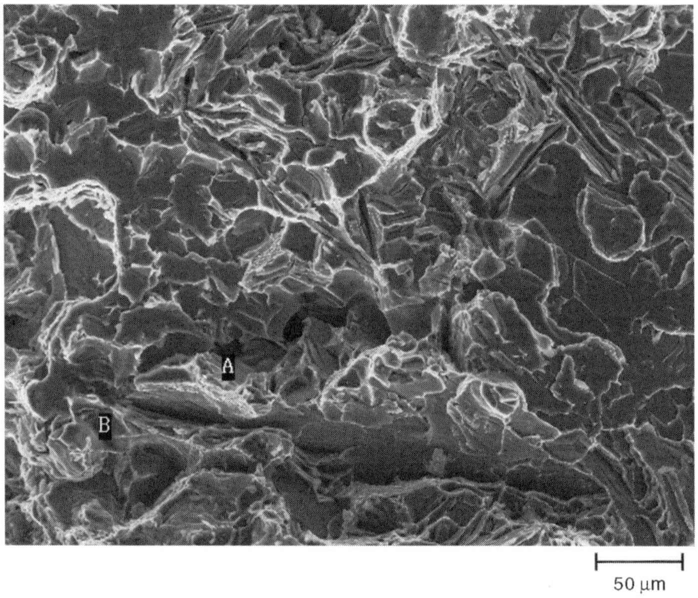

Fig. 7.53 Transcrystalline fracture of medium-developed surface. The crack crossed the cleavage planes in the brittle eutectic phases. Secondary rectilinear cracks are visible in the particles of these phases. 350×

Fig. 7.54 Detail of [A] in Fig. 7.53. The microregions of the matrix were deformed. The dimples in the shear zones and the microvoids on the tear ridge of the micronecks have formed during plastic deformation of the matrix. The band of the waved steps and the deep secondary crack are visible on the cleavage facet in the silicon precipitate (see Fig. 7.14 and 7.17). 1250×

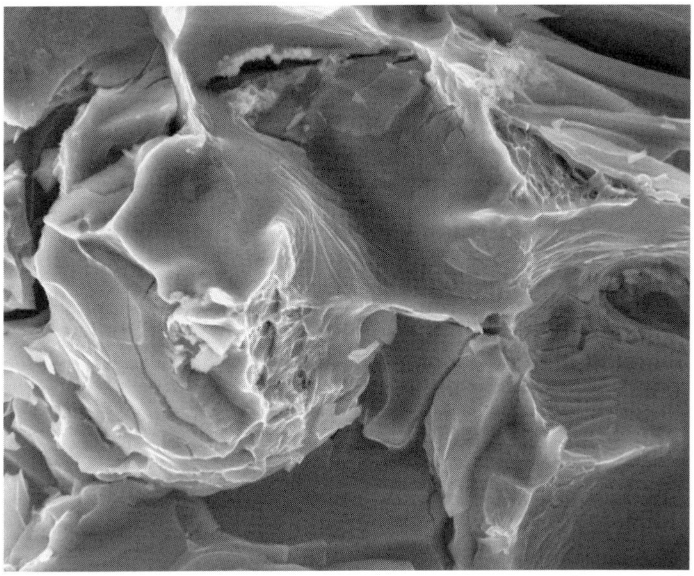

Fig. 7.55 Detail of [B] in Fig. 7.53. The morphology of the waved step bands on the cleavage facets in the silicon particle is characteristic of Wallner's lines. The dimples in the α-aluminum solid solution are a result of plastic deformation. 1500×

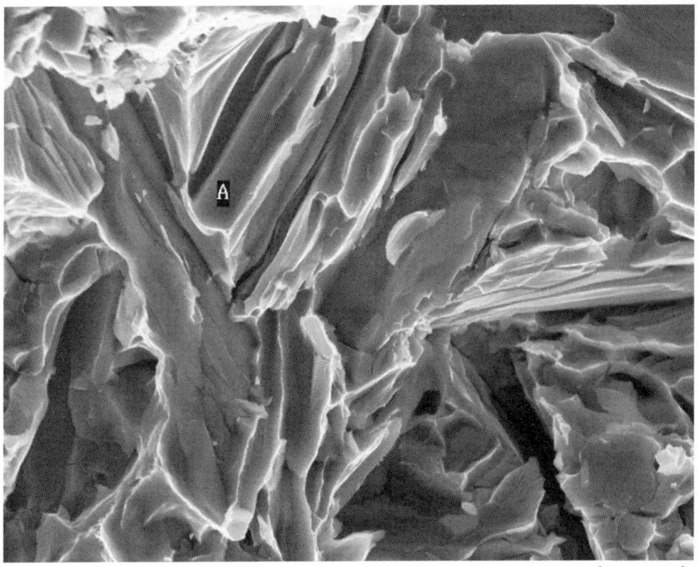

Fig. 7.56 Transcrystalline brittle fracture. In the silicon particles the bands of the parallel cleavage steps and the secondary cracks have formed. Region [A] is detailed in Fig. 7.57. 1300×

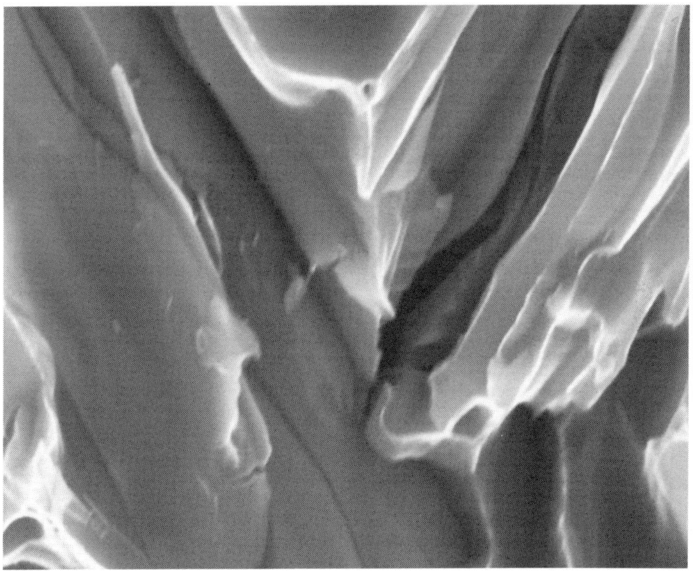

Fig. 7.57 Detail of [A] in Fig. 7.56. The inclined step bands on the cleavage facet form a chevron. Among the steps, the secondary crack was formed. 5000×

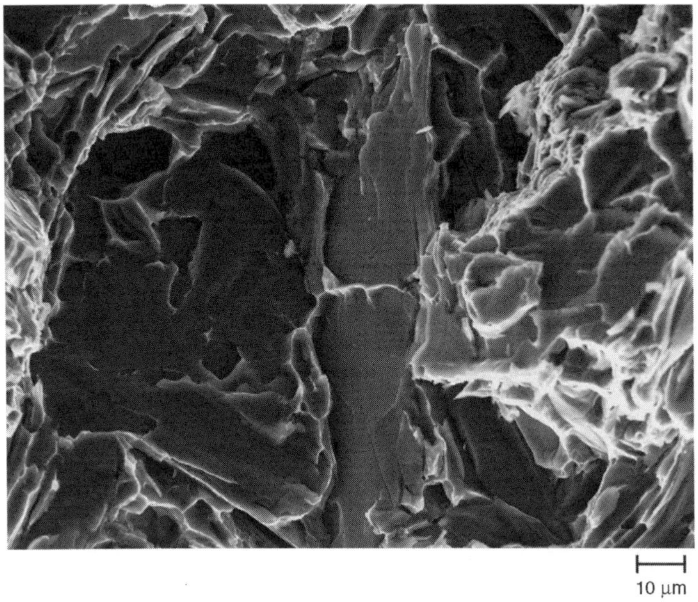

Fig. 7.58 Brittle crack front crosses the precipitate of the β-AlFeSi phase (platelike shape). Two parallel cleavage facets in the β-AlFeSi phase are separated with branched steps. In the matrix area, a trace of plastic deformation, in form of the ductile dimples, is visible. 800×

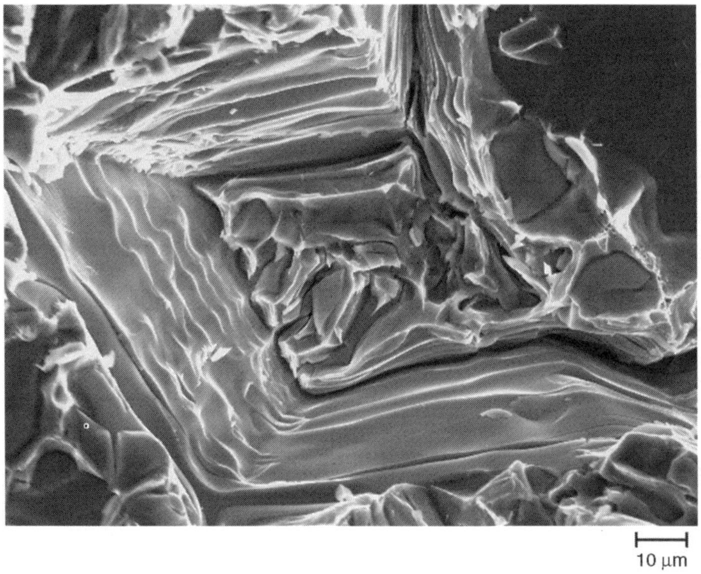

Fig. 7.59 Transcrystalline brittle fracture. The main crack propagated on the cleavage planes of the brittle phase α-Al(FeMn)Si and formed the band of the crossed steps. 650×

CHAPTER 8

Alloy 390.0 (AlSi21CuNi)

Composition and Properties

Alloy	Designation	Chemical composition (main components), wt%						Properties(a)	
		Si	Cu	Mg	Mn	Ni	Fe	R_m, MPa	A_5, %
As-examined	AlSi21CuNi	20.0–23.0	1.1–1.5	0.5–0.9	0.1–0.3	0.8–1.1	0.6 max	200	0.2
Aluminum Association standard(b)	390.0	16.0–18.0	4.0–5.0	0.45–0.65	0.10 max	. . .	1.3 max	200	. . .

(a) R_m, ultimate tensile strength; A_5, elongation measured over a length of 5.65 S_o, where S_o is the cross-sectional area of the test specimen before the test. (b) Alloy 390.0 is registered with the Aluminum Association and designated by ASTM for die casting. R_m listed is a typical value for a separately cast specimen of F or T5 temper and is not specified.

Microstructures

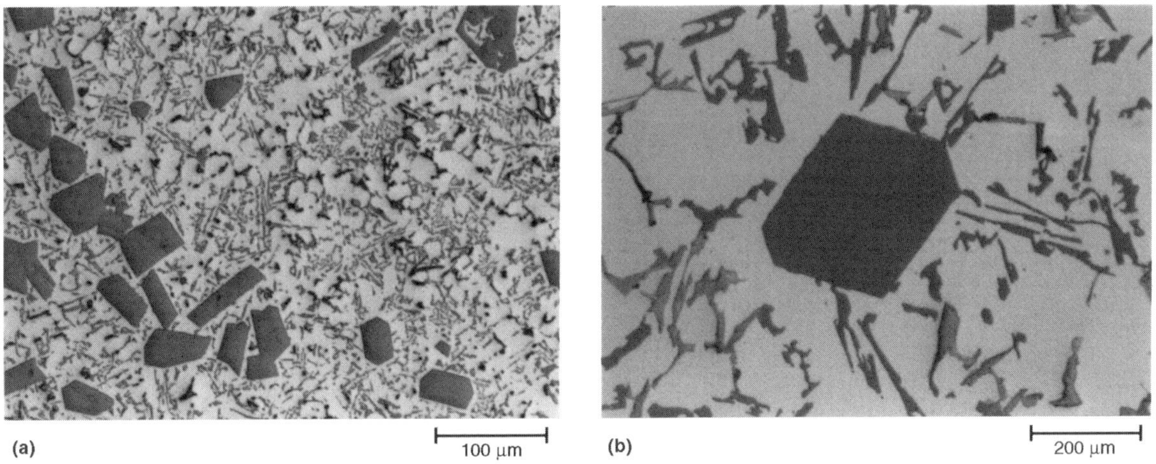

Fig. 8.1 Microstructures of AlSi21CuNi (Alloy 390.0). Light microscope micrographs; etched with 1% HF. (a) As-cast modified, 150×. (b) As-cast modified, 750×

Fracture Profiles of Alloy 390.0 (AlSi21CuNi), Refined, Modified, Permanent Mold Casting

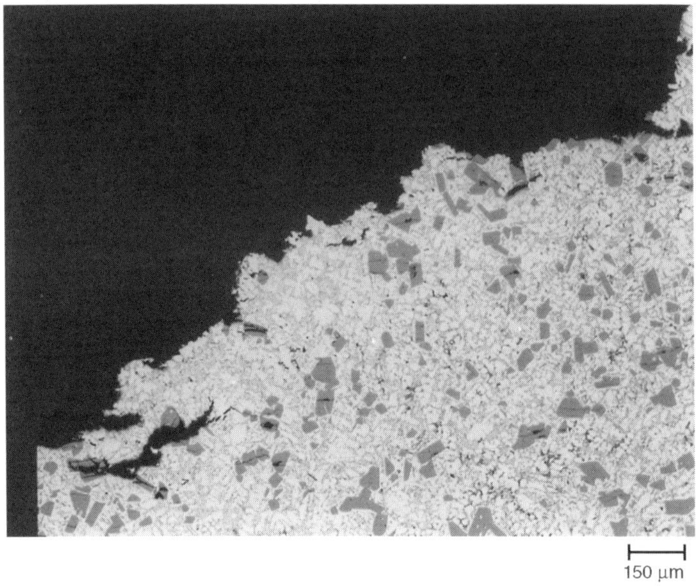

Fig. 8.2 Fracture profile of specimen after static tensile test. Both secondary and internal cracks in primary silicon particles are visible. 50×

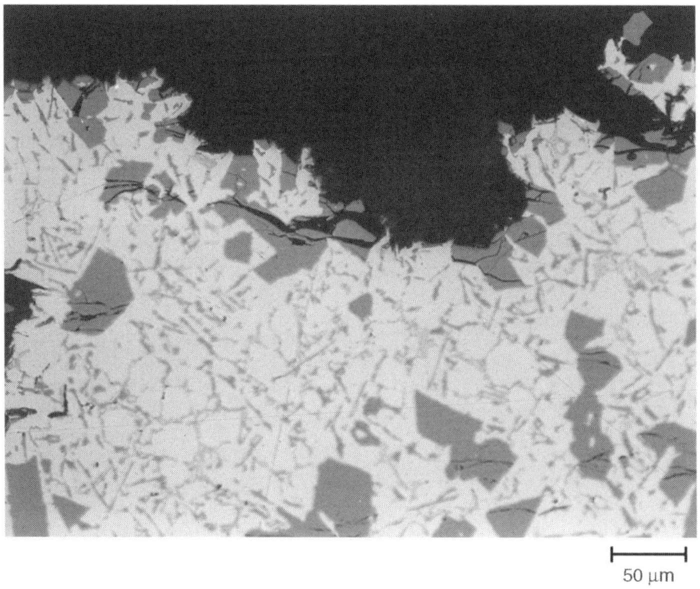

Fig. 8.3 Fracture profile of specimen after static tensile test. The main crack front passes by the successive cleavage planes of the primary silicon particles. Among the silicon precipitates are the sharp micronecks of the deformed α-aluminum solid solution. In the subsurface zone, the numerous secondary cracks of the primary silicon particles are visible. 200×

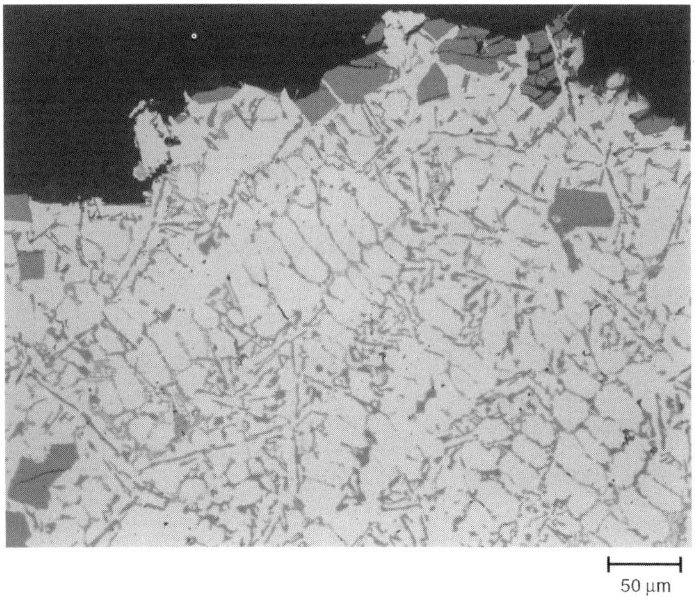

Fig. 8.4 Fracture profile of specimen after static tensile test. The main crack crossed the eutectic region and the parallel cleavage planes in the primary silicon precipitates. In some of them, the branched secondary cracks are visible. Among the silicon precipitates the deformed microligaments of the α-aluminum solid solution have cracked. 200×

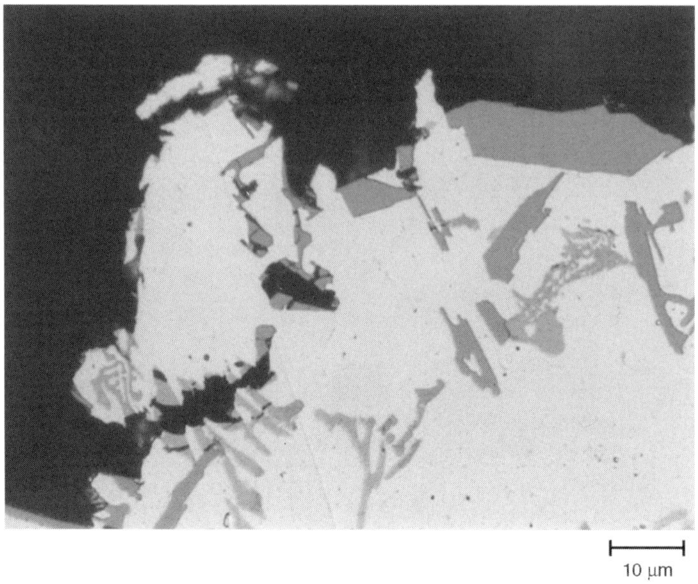

Fig. 8.5 Detail of the profile visible in Fig. 8.4. Sharp, short micronecks of the deformed α-aluminum solid solution are visible. In the primary silicon precipitates and in the brittle eutectic phases, both secondary and internal cracks are present. 1000×

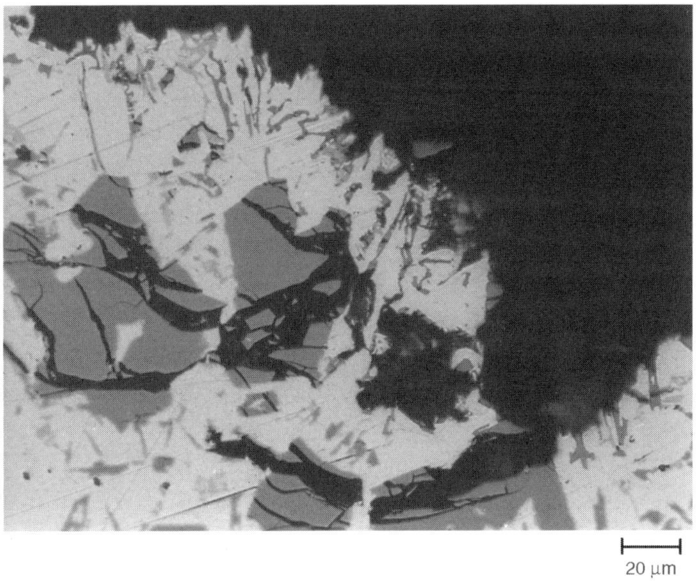

Fig. 8.6 Fracture profile of specimen after static tensile test. The main crack crossed the eutectic region. Among the eutectic silicon particles, the short, sharp micronecks of the deformed α-aluminum solid solution have cracked. In the primary silicon precipitates and the brittle eutectic phases the numerous, branched, secondary cracks were formed. 400×

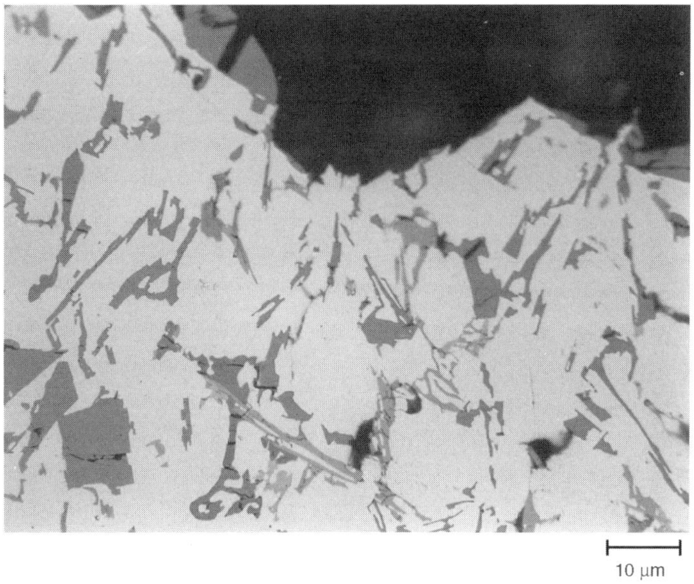

Fig. 8.7 Fracture profile of specimen after static tensile test. The main crack crossed the eutectic region between two primary silicon precipitates. It has propagated on the cleavage planes in primary and eutectic silicon crystals. The short, sharp micronecks of the α-aluminum solid solution among the eutectic silicon particles are present. 1000×

Fracture Surfaces of Alloy 390.0 (AlSi21CuNi), Refined, Modified, Permanent Mold Casting, Fracture after Static Tensile Test

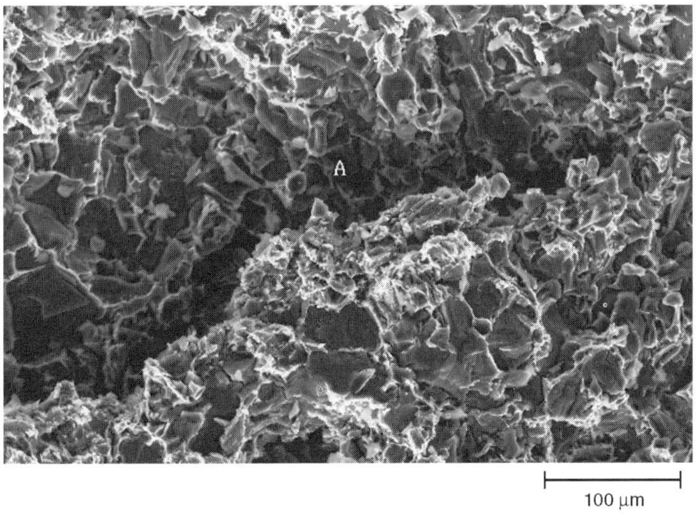

Fig. 8.8 Transcrystalline fracture of greatly developed surface. The crack crossed the cleavage planes of the primary silicon crystals and the regions of a eutectic α-aluminum + silicon. The deep secondary crack also is visible (see Fig. 8.4). 180×

Fig. 8.9 Detail of [A] in Fig. 8.8. Edge of the secondary crack. Brittle cracks in the eutectic phases are visible. 950×

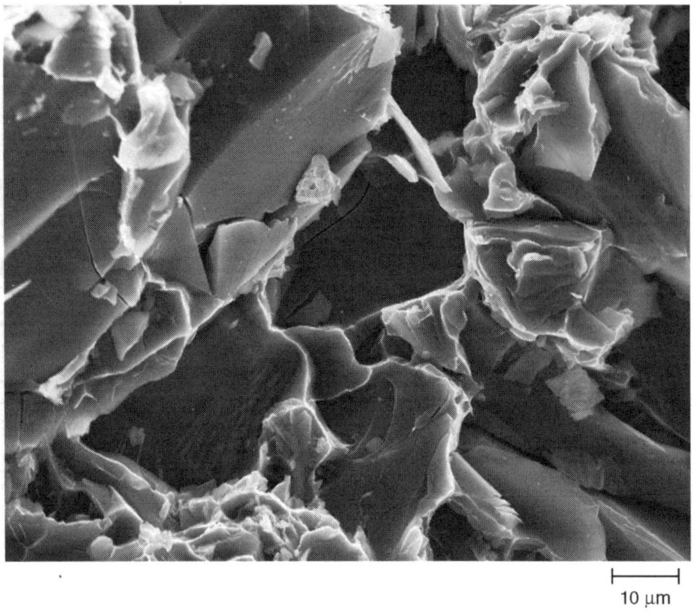

Fig. 8.10 Transcrystalline fracture in the eutectic zone. Steps have formed on the cleavage facets of the brittle eutectic phases. Bands of very small dimples are visible in the deformed regions of the α-aluminum solid solution. 950×

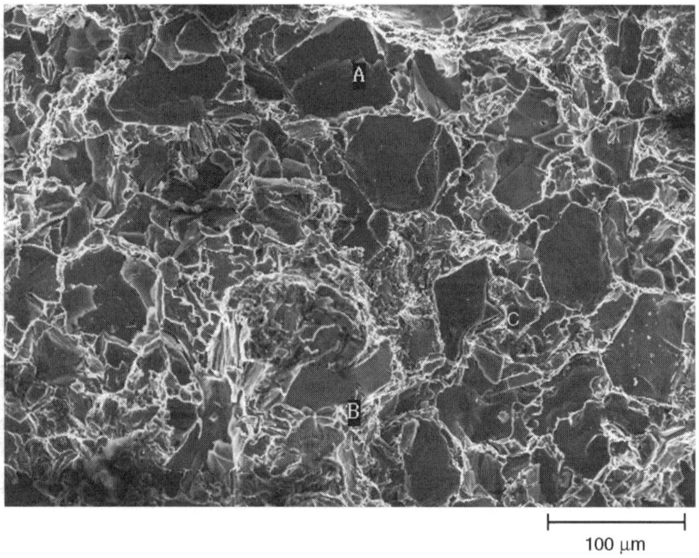

Fig. 8.11 Transcrystalline fracture of weakly developed surface. The crack front crossed the cleavage planes of the primary silicon crystals and the eutectic α-aluminum + silicon regions. (See Fig. 8.4.) 200×

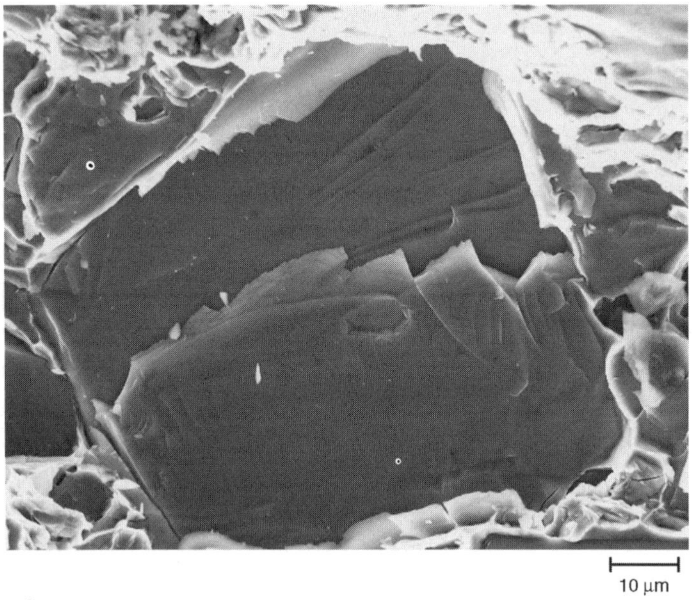

Fig. 8.12 Detail of [A] in Fig. 8.11. Cleavage fracture in the primary silicon crystal. The steps between two parallel cleavage facets are visible (see Fig. 8.5). They form river patterns (in top of micrograph). The direction of the joining of the cleavage steps points out the direction of the crack front propagation. 1000×

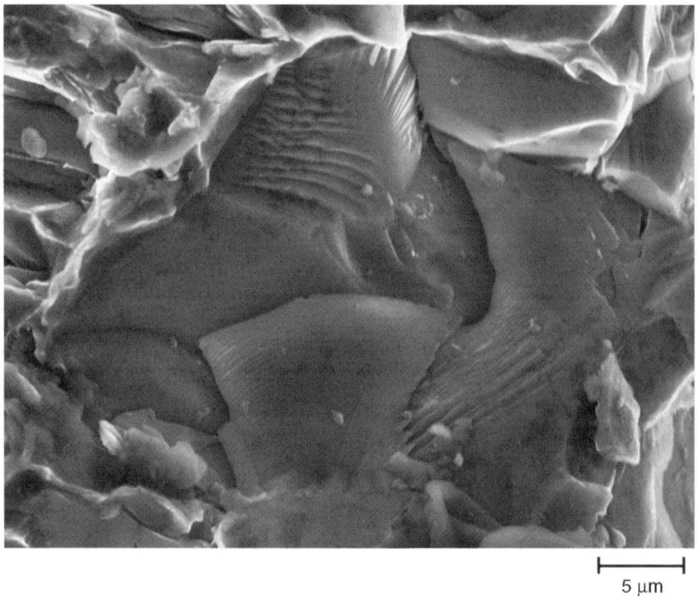

Fig. 8.13 Detail of [B] in Fig. 8.11. The developed cleavage crack in the primary silicon crystal is visible. The steps and Wallner's lines (in shape of the waved bands) can be observed on the cleavage facets. 2500×

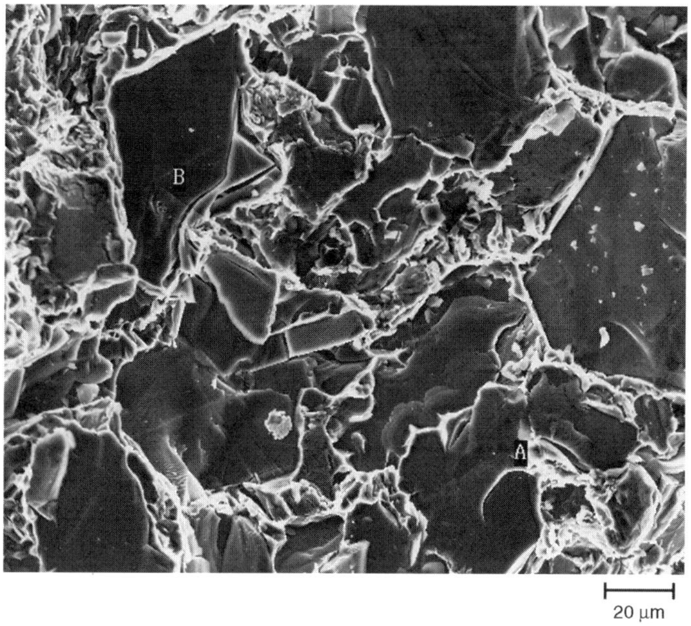

Fig. 8.14 Detail of [C] in Fig. 8.11. The cracks in several primary silicon crystals crossed the cleavage planes, almost parallel to the average fracture plane in this microregion. The dimples resulted from local plastic deformation of the α-aluminum solid solution. 500×

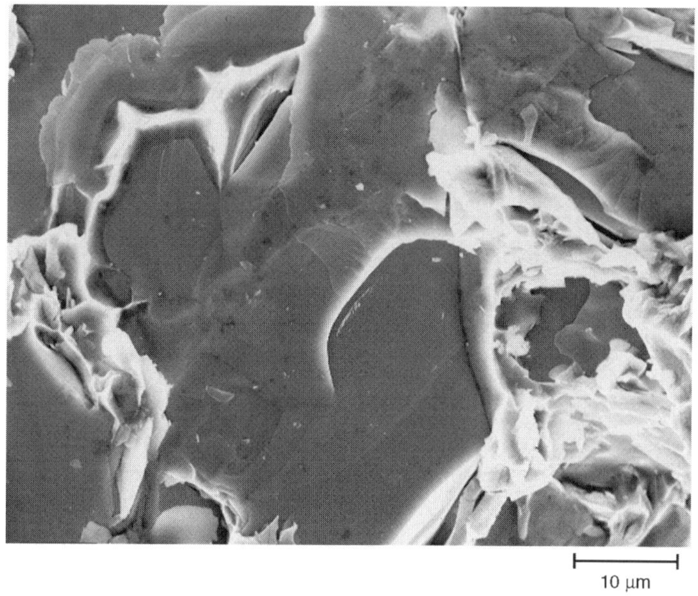

Fig. 8.15 Detail of [A] in Fig. 8.14. The screw step on the cleavage facet of the primary silicon crystal was formed after the crack front was crossed at the low-angle screw boundary. The α-aluminum solid solution in the interface was deformed. 1500×

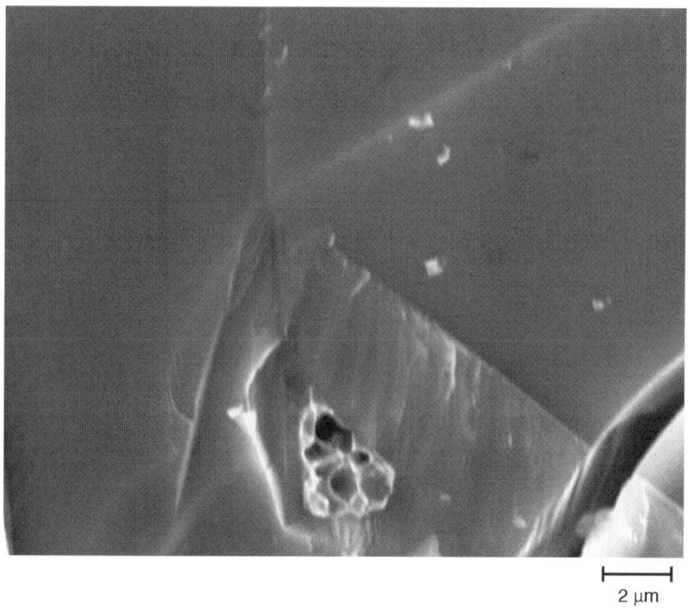

Fig. 8.16 Detail of [B] in Fig. 8.14. Homogeneous decohesion took place on three visible cleavage facets; this is indicated by the absence of the steps. On the fourth plane, cleavage steps and bands of waved Wallner's lines have formed. 5000×

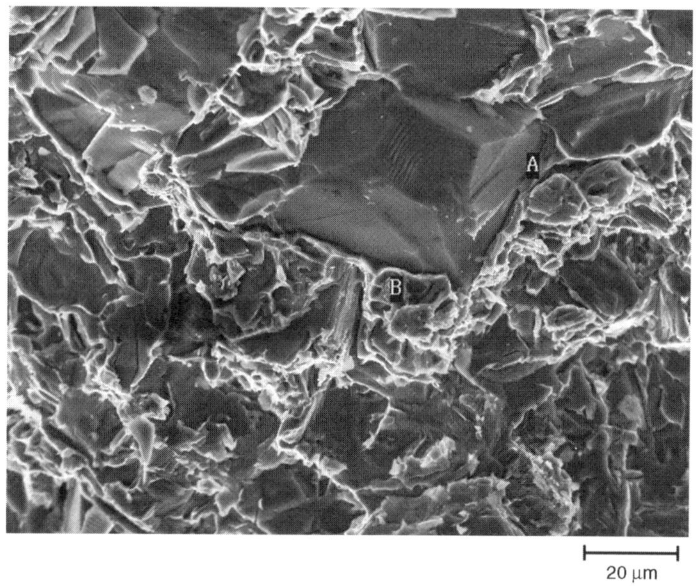

Fig. 8.17 Transcrystalline fracture of mainly brittle character. The crack crossed the large primary silicon crystal, round with eutectic α-Al + Si zone. 700×

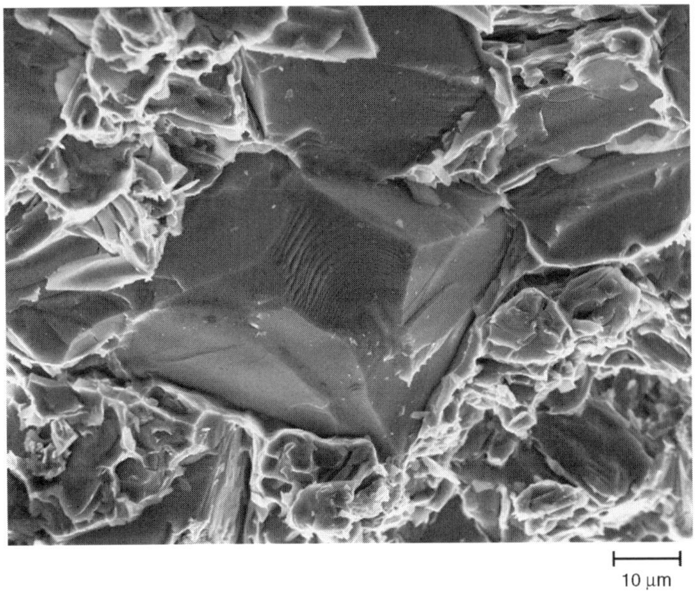

Fig. 8.18 Detail of [A] in Fig. 8.17. Cleavage facets in primary silicon crystal form the part of the hexahedron surface with visible cleavage steps and bands of Wallner's lines. Traces of dimples in the α-aluminum solid solution can be observed. 1000×

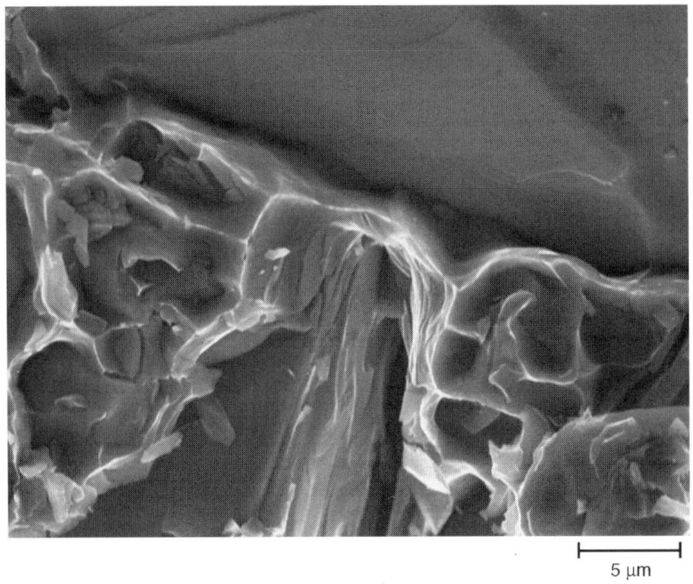

Fig. 8.19 Detail of [B] in Fig. 8.17. Morphology of the interface zone between α-aluminum and silicon. On the interface, the interface cohesion was retained. Single voids can be observed on the edges of the micronecks in the α-aluminum solid solution. 3000×

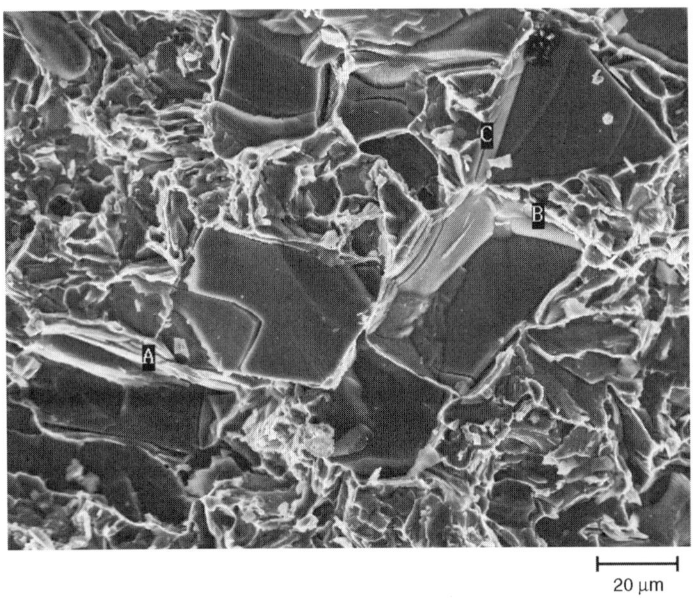

Fig. 8.20 Transcrystalline fracture of mainly brittle character. The crack crossed the cleavage planes in the primary silicon crystal round with the eutectic α-aluminum + silicon zone. The start of decohesion can be observed on the interface between primary silicon and α-aluminum solid solution. 600×

Fig. 8.21 Detail of [A] in Fig. 8.20. The crack crossed the cleavage planes in the primary silicon crystal. The bands of the parallel steps and rivers form the river patterns. They are an effect of the crack front displacement on the successive cleavage planes. 2500×

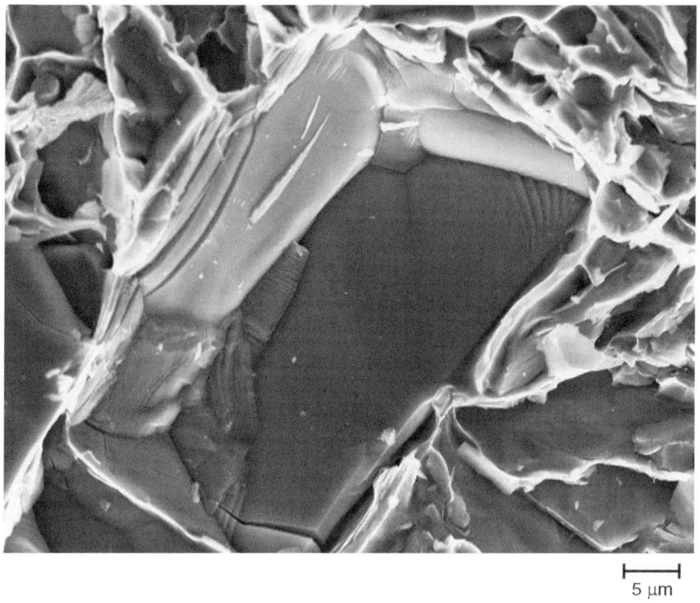

Fig. 8.22 Detail of [B] in Fig. 8.20. The crack crossed the several cleavage planes of different orientation in the primary silicon crystal. Secondary cracks have also formed in the silicon crystal. The decohesion zone on the interface between primary silicon and α-aluminum solid solution can be observed. 1500×

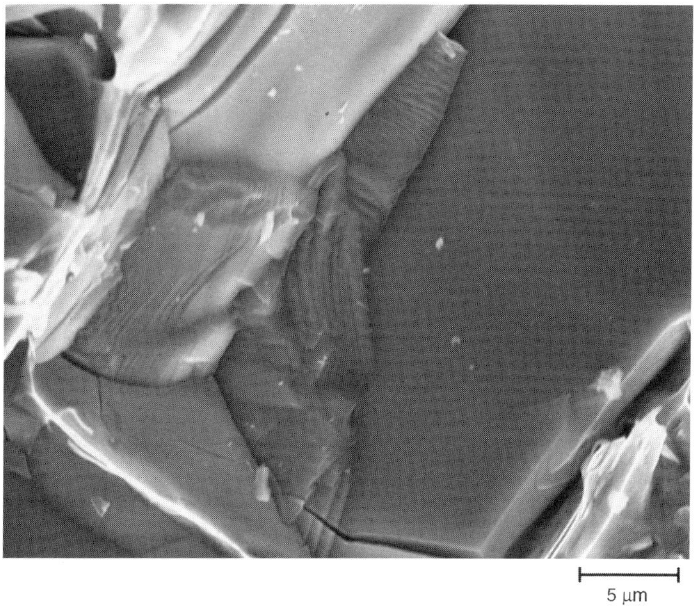

Fig. 8.23 Detail of Fig. 8.22. The homogeneous cleavage as the main decohesion mechanism in the primary silicon crystal is indicated by the absence of the steps. Nevertheless, microdeformation traces are visible on some cleavage facets. Bands of parallel cracks and Wallner's lines can be observed (left side of the micrograph). 2600×

Fig. 8.24 Detail of [C] in Fig. 8.20. The bands of the cleavage steps of different height were formed by displacement of the crack front on the successive cleavage planes in the silicon crystal. 6500×

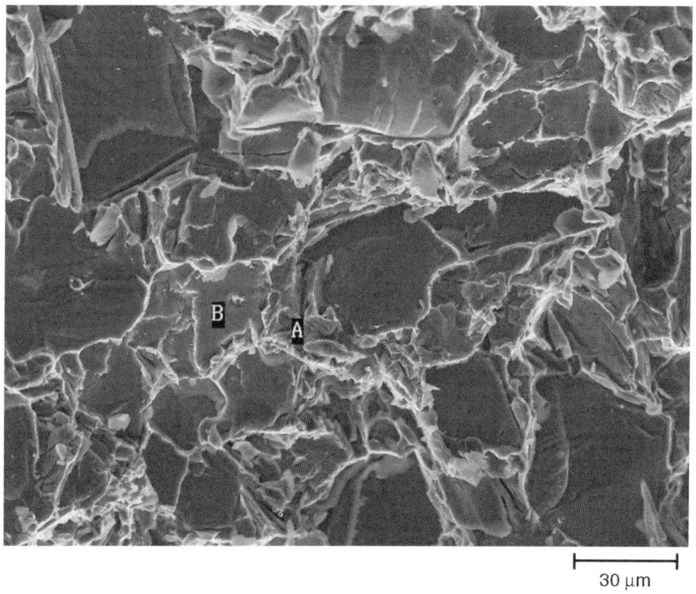

Fig. 8.25 Transcrystalline fracture of mainly brittle character of weakly developed surface. The crack crossed the cleavage planes of the primary silicon crystal and the eutectic α-Al + Si. 500×

Fig. 8.26 Detail of [A] in Fig. 8.25. The crack crossed the eutectic α-Al + Si (see Fig. 8.6). In the eutectic silicon precipitates, cleavage cracks are present. The microneck of the α-aluminum solid solution was fractured along the tear ridge. In the α-aluminum solid solution the oval, single voids are present. 3200×

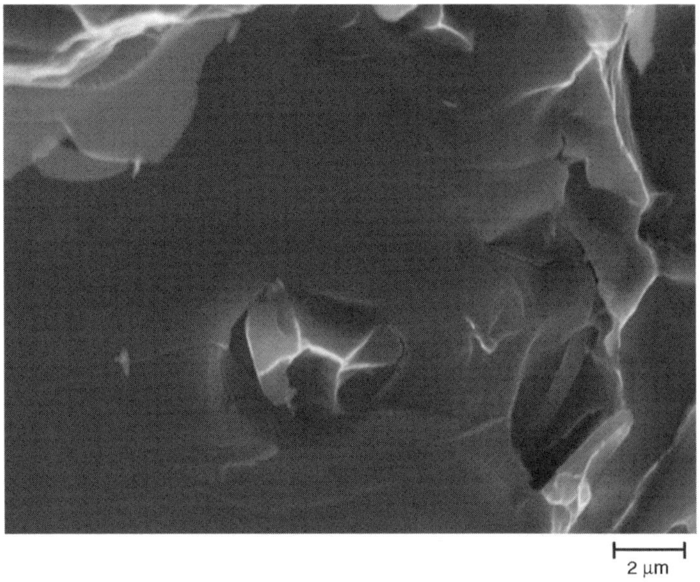

Fig. 8.27 Detail of [B] in Fig. 8.25. Cleavage facets in the silicon crystal. The small cleavage steps are present in the crack initiation zone. 5000×

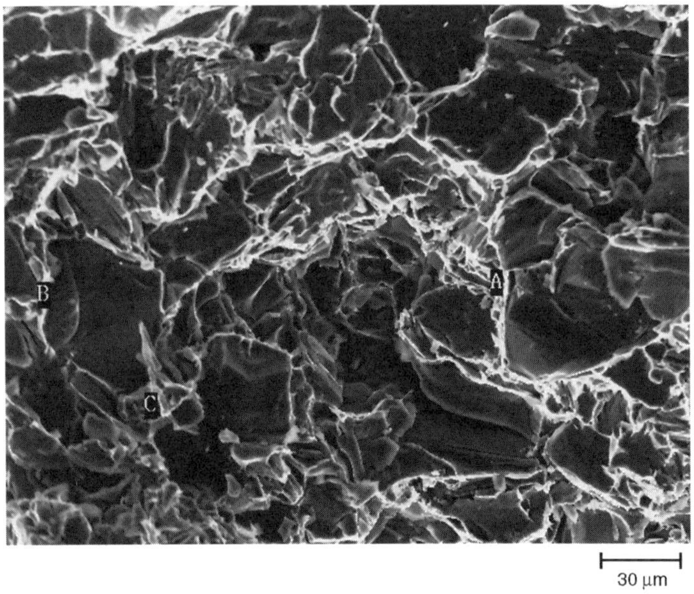

Fig. 8.28 Transcrystalline fracture of medium-developed surface. The several primary silicon crystals, fractured on the cleavage planes, are visible. 400×

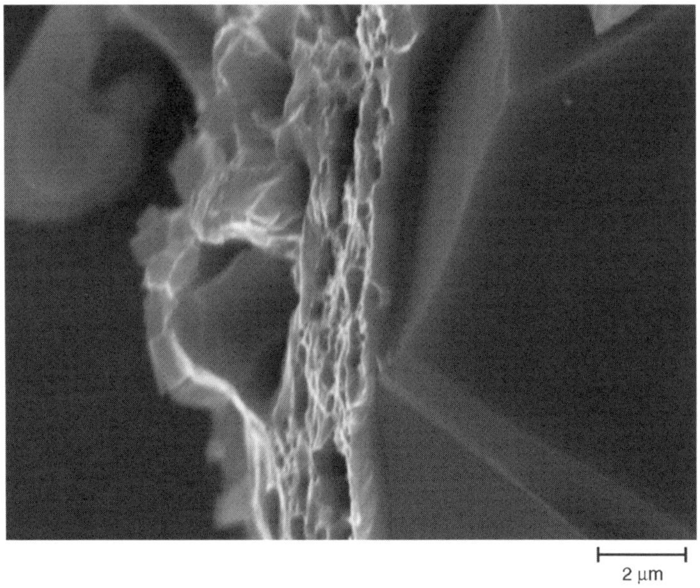

Fig. 8.29 Detail of [A] in Fig. 8.28. The crack front passed in the primary silicon crystals by the several cleavage planes, without forming visible steps. In the α-aluminum solid solution, on the interface between α-aluminum and silicon, bands of oval and equiaxial dimples have formed. The zone of the retained cohesion is present on the interface between α-aluminum and silicon. 5000×

Fig. 8.30 Detail of [B] in Fig. 8.28. Cleavage facets in the silicon crystal. The step bands form the river patterns. The joining of the steps reflects the tendency to decrease the surface energy. The step displacement on the successive cleavage planes took place. After the secondary crack formation two parts of the crystal dislocated relatively. 7500×

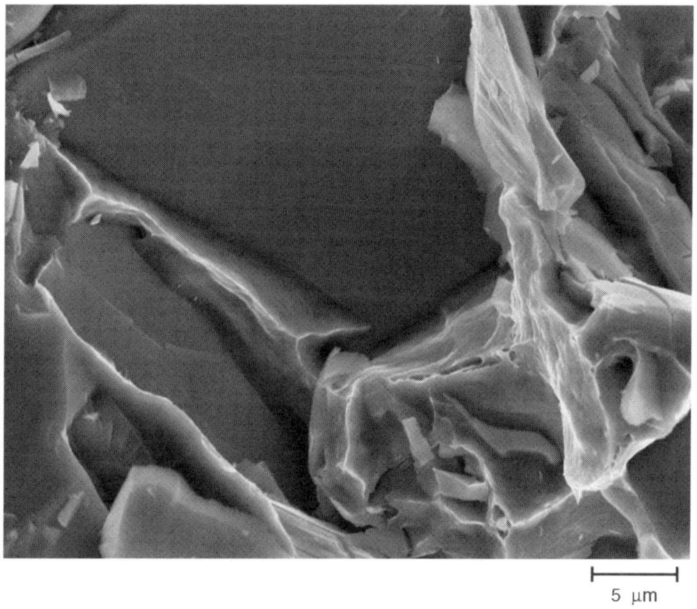

Fig. 8.31 Detail of [C] in Fig. 8.28. The deformed α-aluminum solid-solution zone between primary and eutectic silicon crystals was fractured along the tear ridges. Voids and dimples, formed after their coalescence, can be observed on the tear edges along the line of their ductile crack. 2500×

Fig. 8.32 Two cells formed in α-aluminum solid solution around the cracked silicon precipitates. The zones of the deformed α-aluminum solid solution were fractured in the plastic manner. In the boundary of the deformed zones, the voids coalesce, and the first stage of dimple formation can be observed. On the interfaces (top of the micrograph) the interface cohesion was retained. 3600×

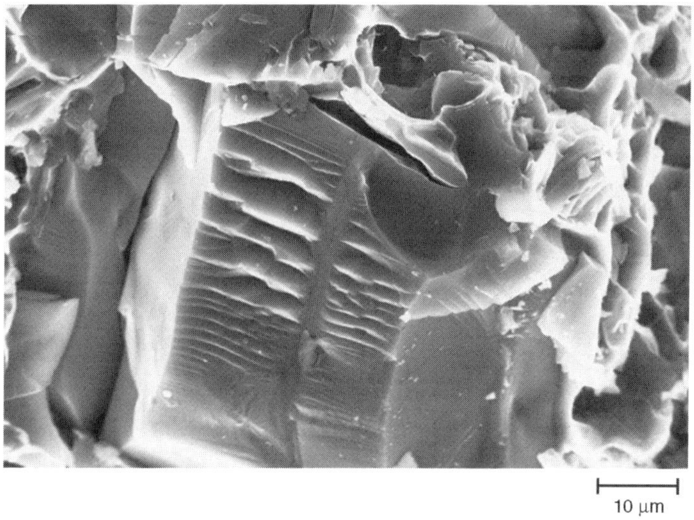

Fig. 8.33 Bands of parallel cleavage steps in the primary silicon crystal. 1200×

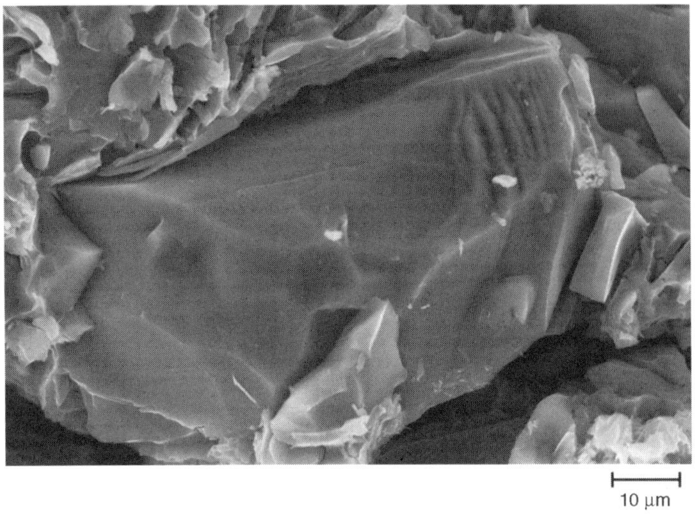

Fig. 8.34 Fracture surface in the primary silicon crystal. The crack crossed the several cleavage planes of different orientation. A band of Wallner's lines is visible. 1000×

Fig. 8.35 Step bands resulting from the front crack displacement from one cleavage plane to another. The trace of slip can be observed (right side of the micrograph). 7000×

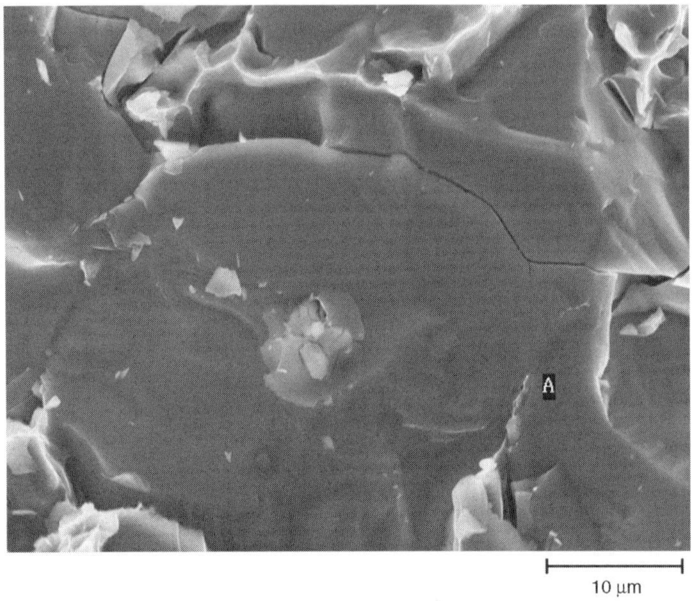

Fig. 8.36 Cleavage facet in the primary silicon crystal; the crack initiation zone is visible. 2000×

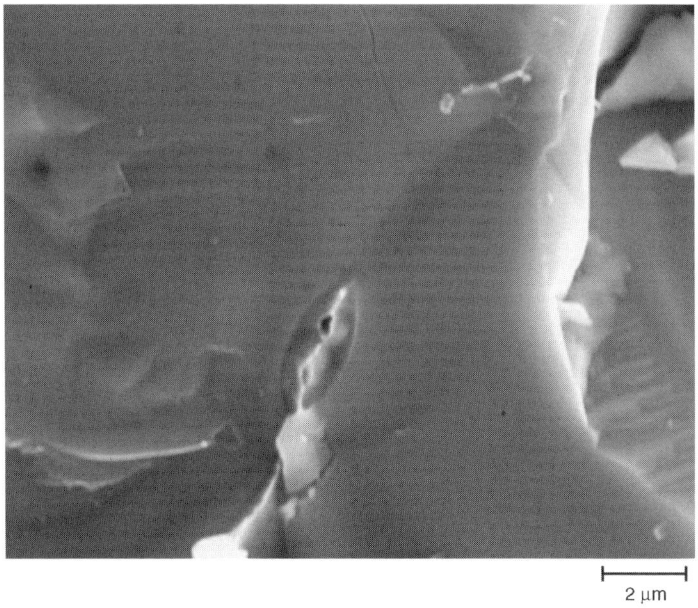

Fig. 8.37 Detail of [A] in Fig. 8.36. Cleavage steps were formed among the parallel cleavage planes. Voids are present in the α-aluminum solid solution zone. 6000×

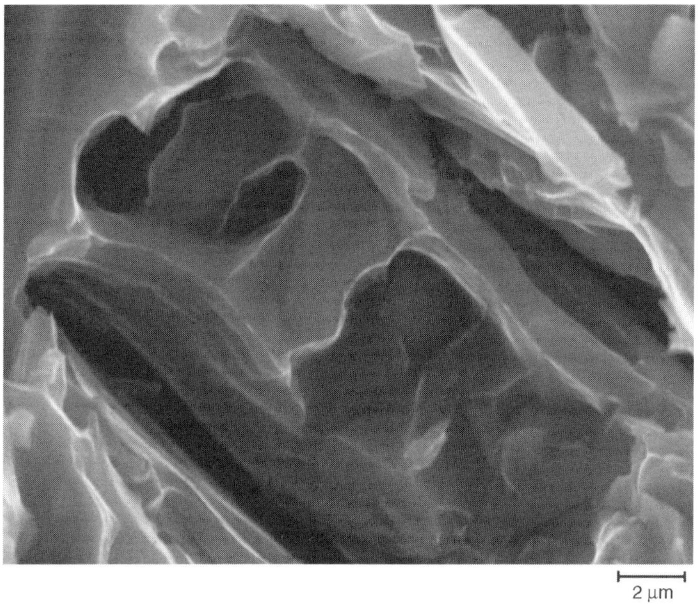

Fig. 8.38 Fracture surface in the polyphase region α-Al + Si + Al_2Cu. The precipitates of the Al_2Cu phase are situated on the bottom of the shallow cells, formed in deformed solid solution α-aluminum. Secondary cracks are present on the interfaces. 4500×

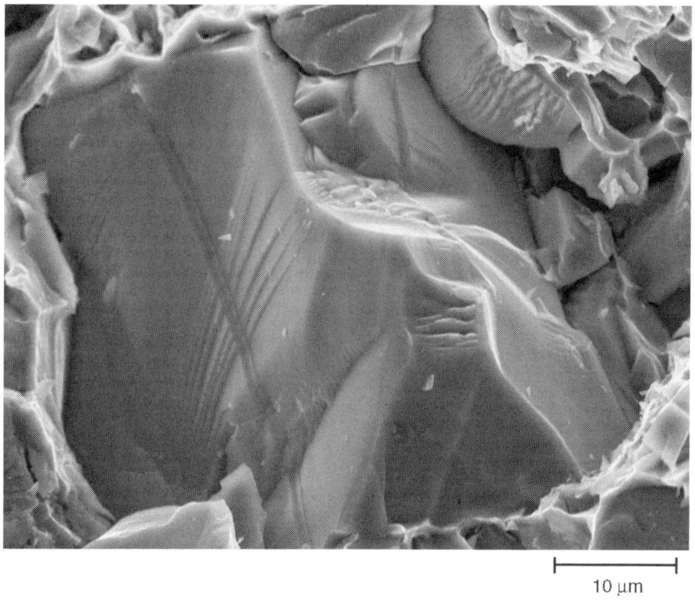

Fig. 8.39 The crack in the primary silicon crystal crossed the cleavage planes. Steps and Wallner's lines are visible. In the α-aluminum solid solution, in the interface zone, plastic microdeformation took place. The interface cohesion on the interfaces was retained. 1800×

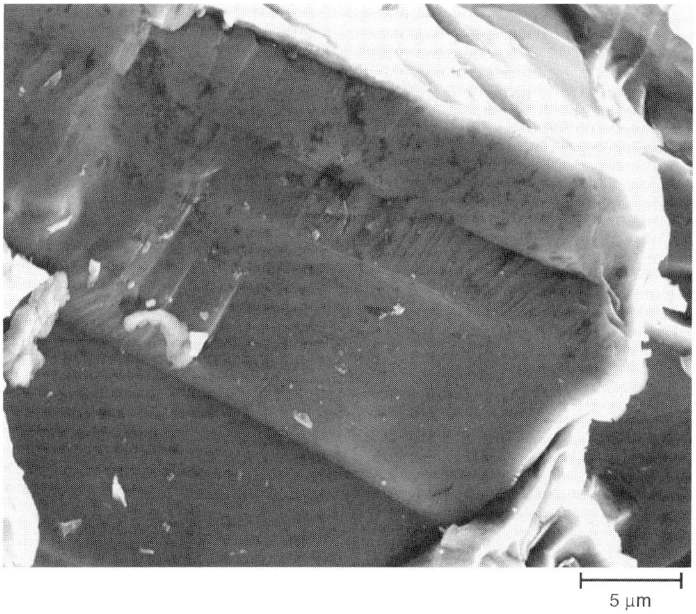

Fig. 8.40 The step system formed on four parallel cleavage planes in the primary silicon crystal. 3000×

CHAPTER 9

Alloy 413.0 (AlSi11)

Composition and Properties

Alloy	Designation	Chemical composition (main components), wt%						Properties(a)	
		Si	Cu	Mg	Mn	Ni	Fe	R_m, MPa	A_5, %
As-examined	AlSi11	10.0–13.0	. . .	. . .	. . .	. . .	1.0 max	200	6
Aluminum Association standard	413.0	11.0–13.0	1.0 max	0.10 max	0.35 max	0.50 max	2.0 max	293	2.5

(a) R_m, ultimate tensile strength; A_5, elongation measured over a length of 5.65 S_o, where S_o is the cross-sectional area of the test specimen before the test. Alloy 413.0 is registered with the Aluminum Association and ASTM for die casting. R_m is a typical value and not specified.

Microstructures

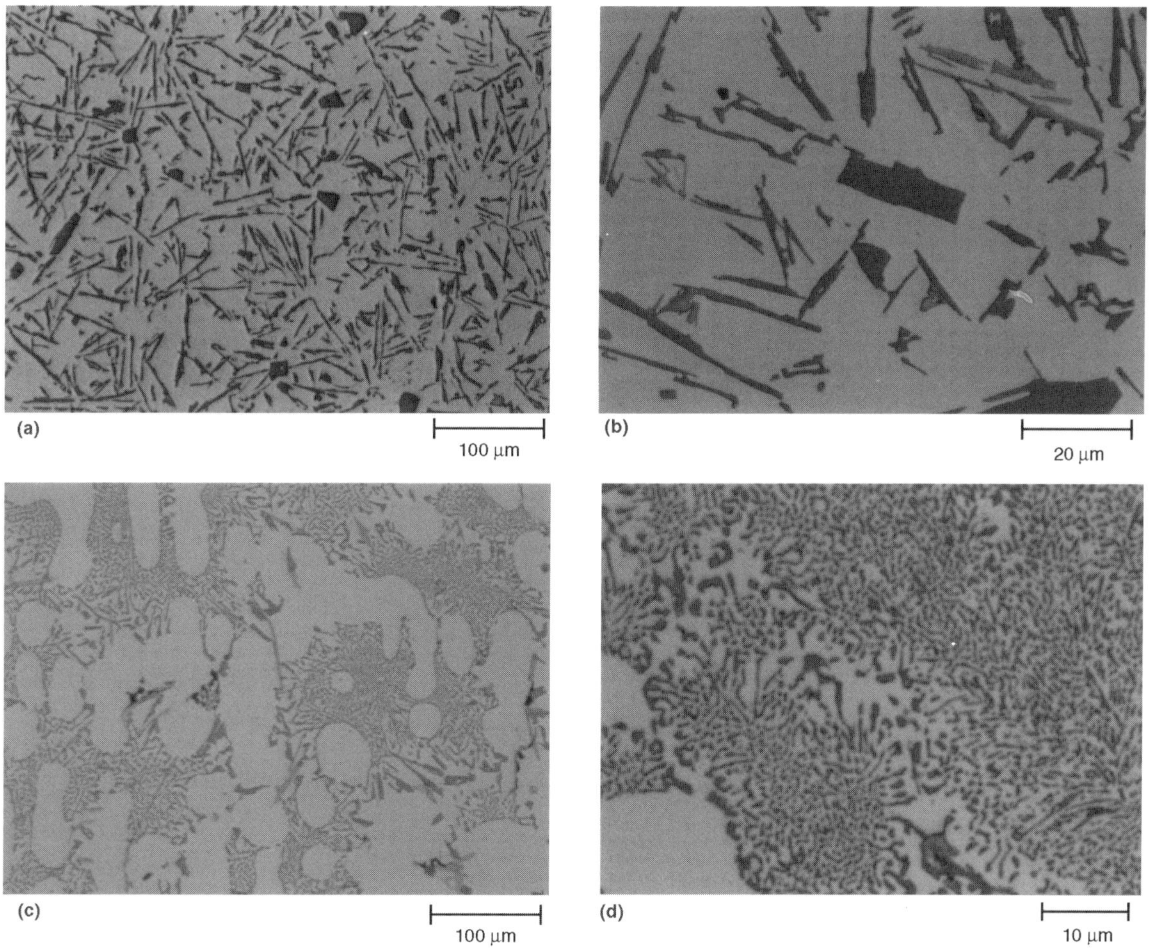

Fig. 9.1 Microstructures of AlSi11 (Alloy 413.0). Light microscope micrographs; etched with 1% HF. (a) As-cast (F), 150×. (b) As-cast (F), 750×. (c) As-cast modified, 150×. (b) As-cast modified, 1200×

Fracture Profiles of Alloy 413.0 (AlSi11), Refined, Modified, Permanent Mold Casting

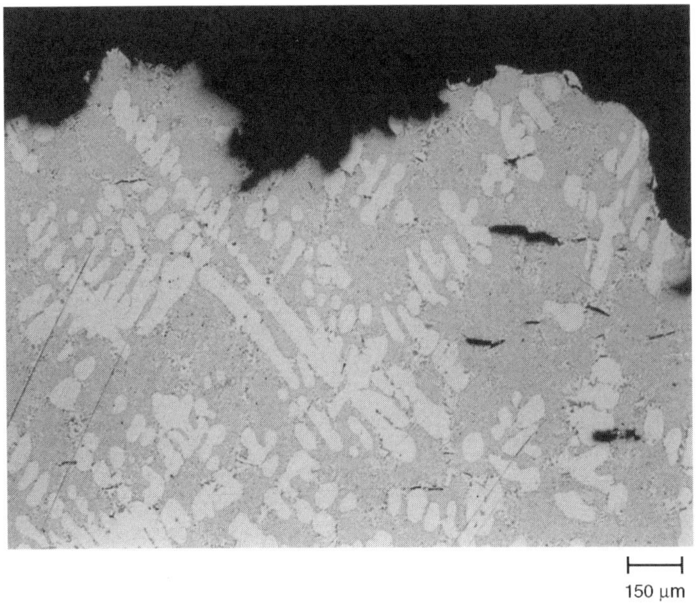

Fig. 9.2 Fracture profile of specimen after static tensile test. The long internal cracks in the eutectic zone (silicon + α-aluminum solid solution) have formed. 50×

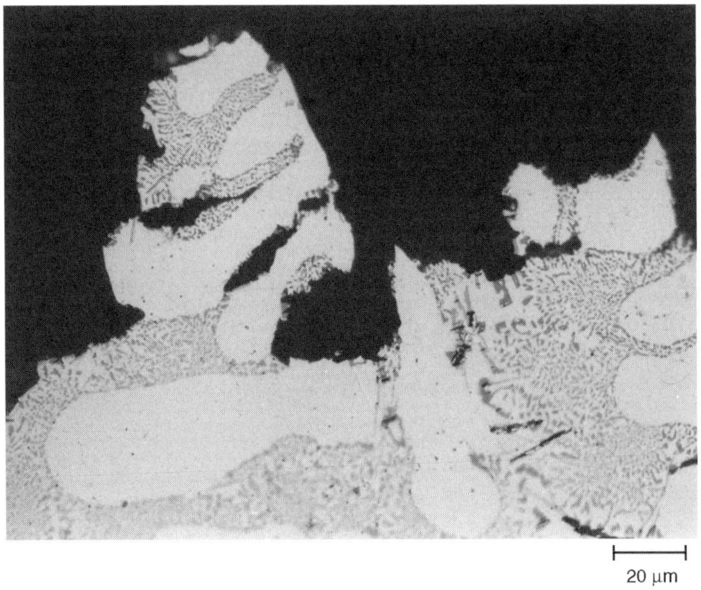

Fig. 9.3 Fracture profile of specimen after static tensile test. The profile of the main crack reflects the morphology of the primary dendritic structure. The main crack crossed the two-phase regions (α-aluminum + silicon) and sheared the dendrite of the α-aluminum solid solution. The micronecks of the deformed α-aluminum solid solution are visible. In the eutectic zones the secondary cracks are visible. 500×

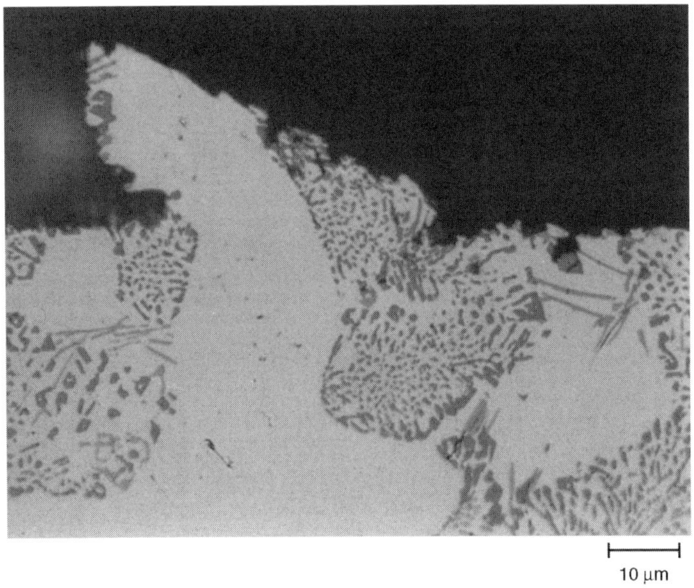

Fig. 9.4 Fracture profile of specimen after static tensile test. The main crack crossed the two-phase zone in the eutectic α-aluminum + silicon. The short microligaments have formed in deformed α-aluminum solid solution, among brittle precipitates of silicon (right side of the micrograph). On the shear edge, in the dendrite of the α-aluminum solid solution, the shear steps are visible. 1000×

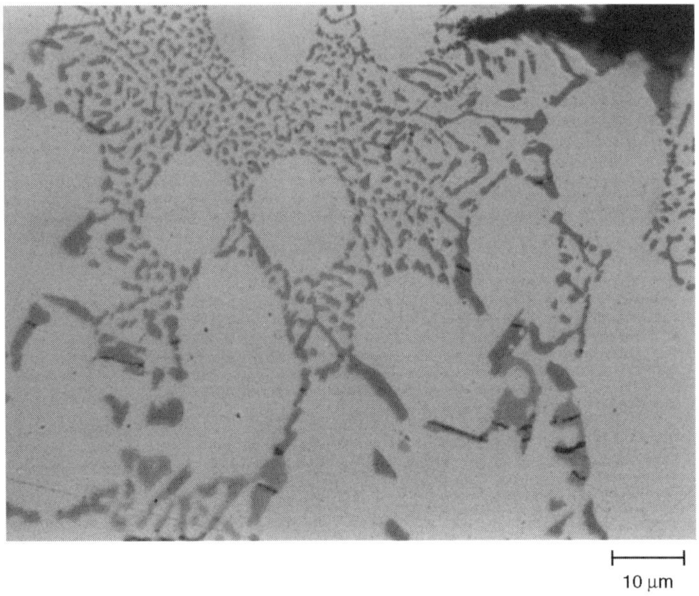

Fig. 9.5 Fracture in specimen after static tensile test. In the two-phase zone, beneath fracture surface, numerous microcracks have formed in the brittle eutectic phases: Si and α-Al(FeMn)Si. 1000×

Fracture Surfaces of Alloy 413.0 (AlSi11), Refined, Permanent Mold Casting, Fracture after Static Tensile Test

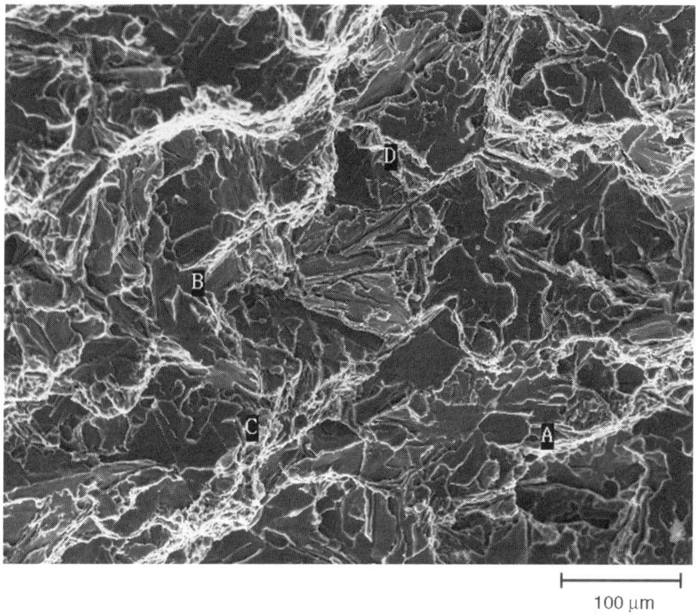

Fig. 9.6 Transcrystalline fracture of weakly developed surface. The crack crossed the cleavage planes of the brittle microstructure constituents. The areas of the fracture of cellular morphology are visible. 180×

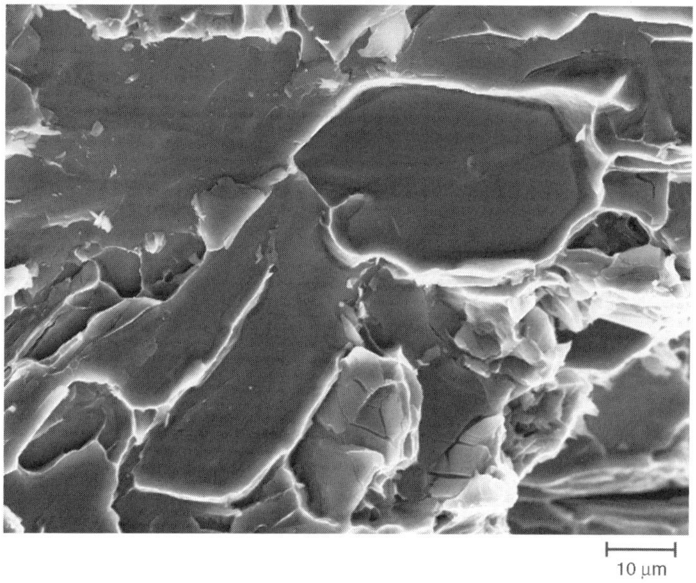

Fig. 9.7 Detail of [A] in Fig. 9.6. Transcrystalline cleavage fracture. The crack crossed the parallel cleavage planes in the silicon particles. The screw and branched steps on these cleavage facets are visible. 1000×

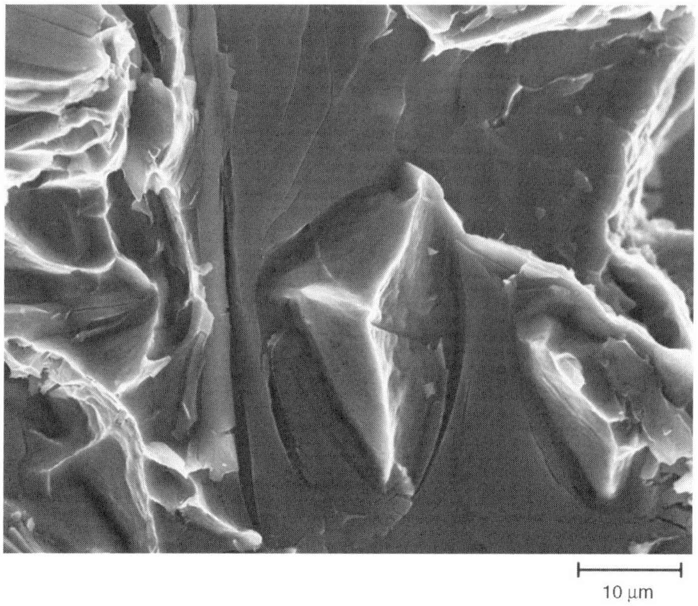

Fig. 9.8 Detail of [B] in Fig. 9.6. On the surface of the cracked silicon precipitate, several cleavage facets, steps, and secondary cracks have formed during main crack front displacement. 1500×

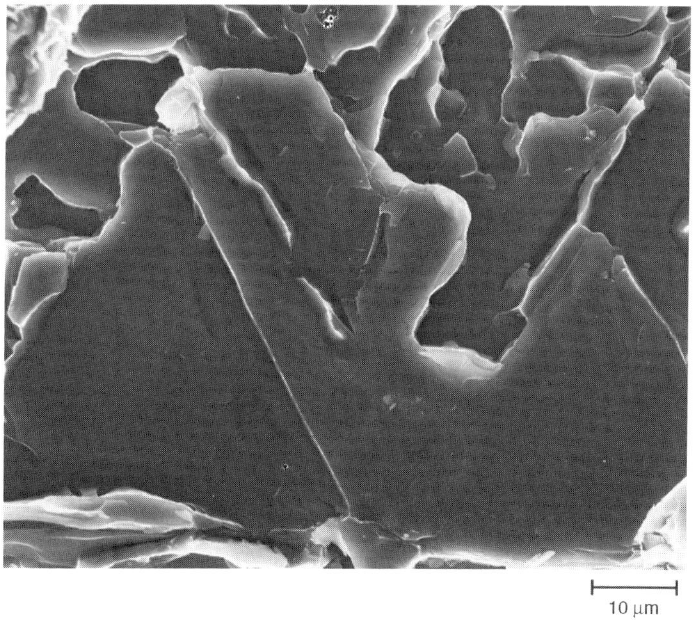

Fig. 9.9 Detail of [C] in Fig. 9.6. Transcrystalline cleavage fracture. On the fracture surface in silicon precipitate, the parallel cleavage planes are separated with the cleavage steps. Rectilinear steps, screw steps, and secondary cracks also have formed in this fracture area. 1200×

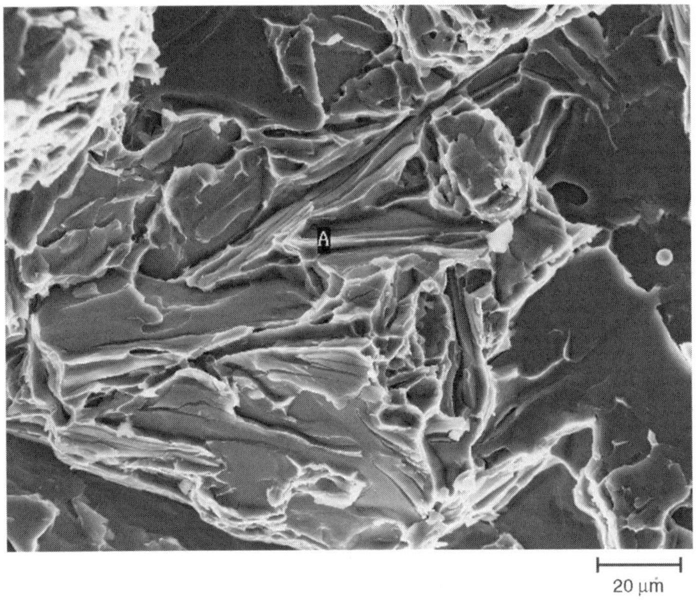

Fig. 9.10 Detail of [D] in Fig. 9.6. Transcrystalline fracture area. The crack front displaced on the successive cleavage planes, of different orientation, in the silicon precipitates. Traces of plastic deformation can be observed in the matrix. 1200×

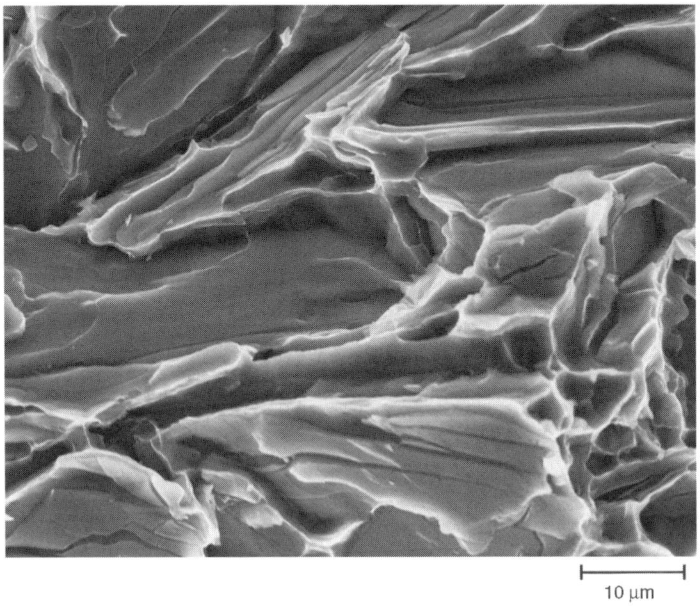

Fig. 9.11 Detail of [A] in Fig. 9.10. Multiplane cleavage crack in the plate-like silicon precipitate. The crack front crossed the successive cleavage planes. The rivers and tongues are the traces of the microdeformation process. Dimples can be observed in the α-aluminum solid solution. 6000×

Fracture Surfaces of Alloy 413.0 (AlSi11), Refined, Modified, Permanent Mold Casting, Fracture after Static Tensile Test

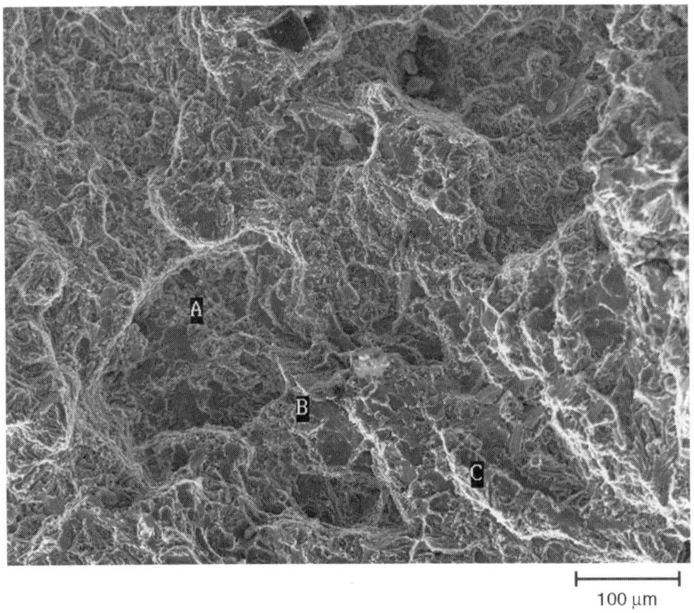

Fig. 9.12 Transcrystalline fracture of greatly developed surface (see Fig. 9.3). The micropores and the nonmetallic inclusion are visible on the fracture surface. 150×

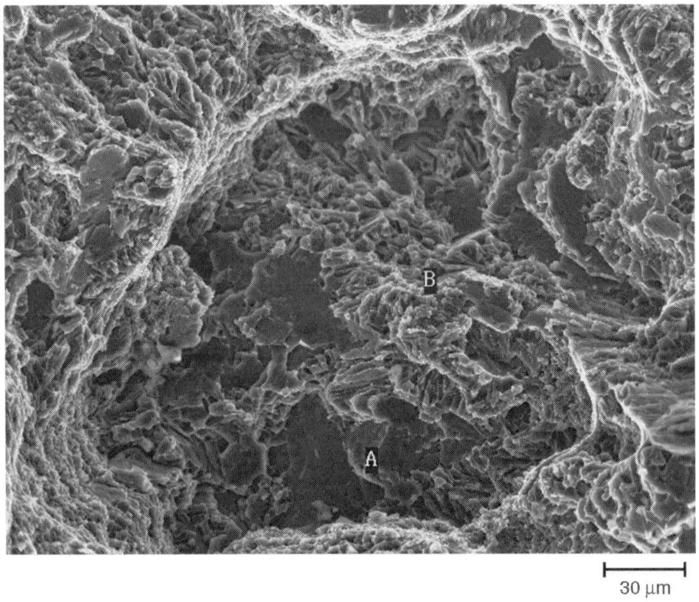

Fig. 9.13 Detail of [A] in Fig. 9.12. The main crack was formed in the eutectic grain. The crack crossed both eutectic silicon precipitates and cellular zones. The fractured eutectic grain is rounded with a deformed α-aluminum solid solution zone. 400×

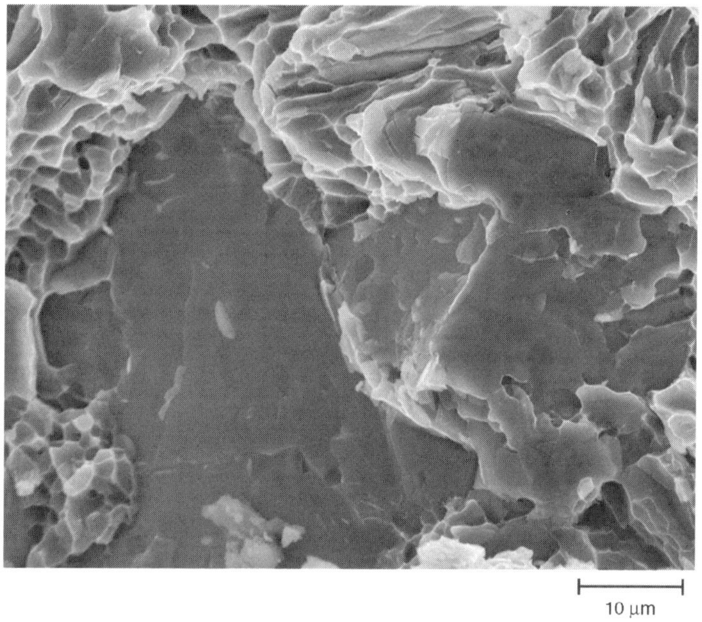

Fig. 9.14 Detail of [A] in 9.13. The crack crossed the parallel cleavage planes in neighboring eutectic silicon precipitates. The steps and tongues are present in the microdeformation zone. 2000×

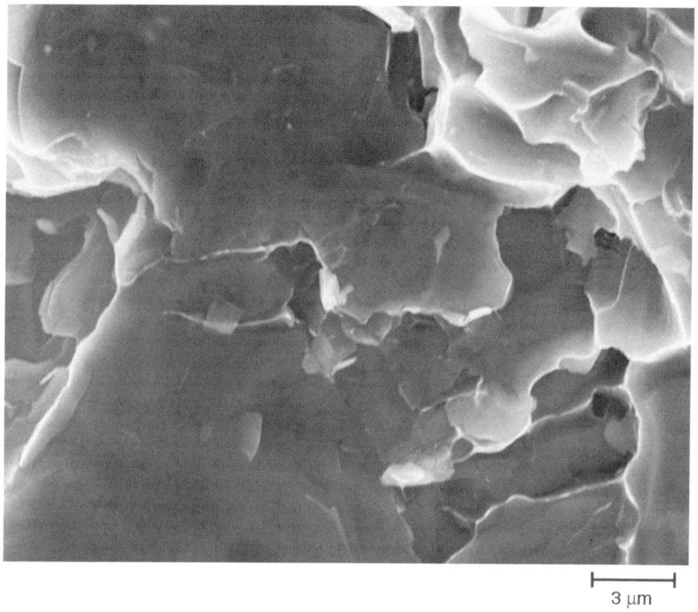

Fig. 9.15 Detail of Fig. 9.14. Steps, rivers, and tongues on the cleavage facets—effects of the microdeformation of the crystal lattice in the eutectic silicon precipitate. 4000×

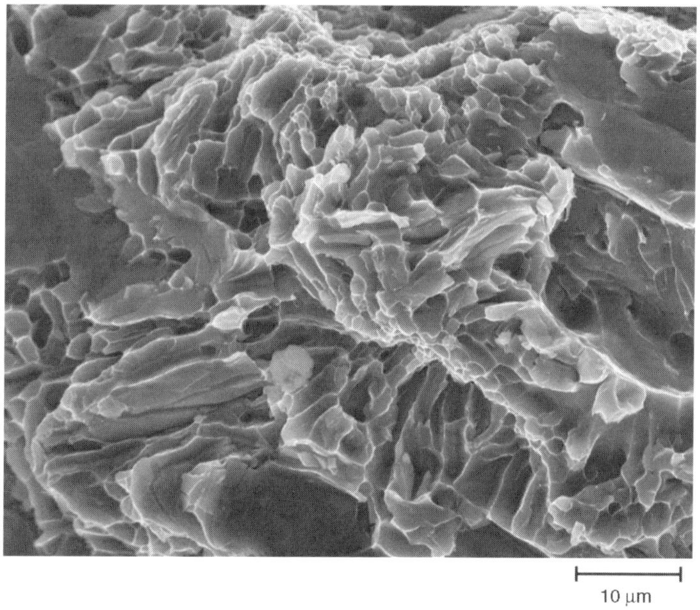

Fig. 9.16 Detail of [B] in Fig. 9.13. Transcrystalline fracture of characteristic, cellular morphology. The main crack displaced in the two-phase region α-Al + Si. 1500×

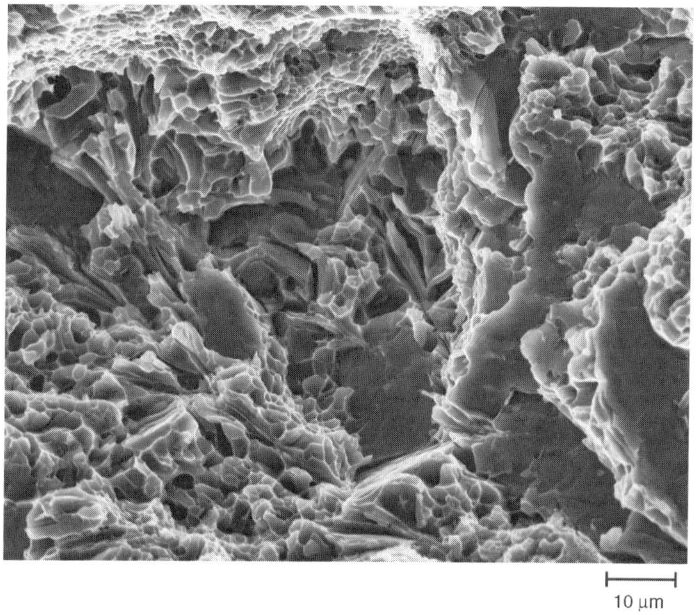

Fig. 9.17 Detail of [B] in Fig. 9.12. The terraces of the cleavage steps have formed during displacement of the crack front on the successive parallel cleavage planes in the silicon precipitates. In the matrix zone the equiaxial and open dimples can be observed. 1000×

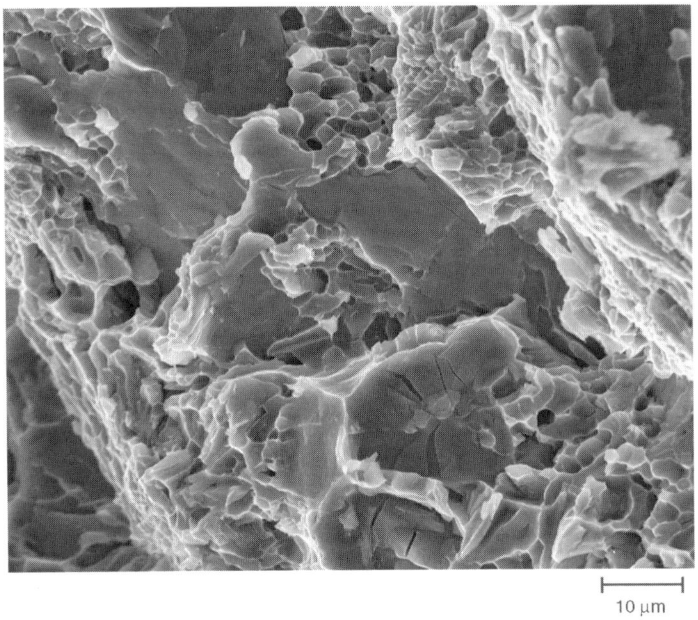

Fig. 9.18 Detail of [C] in Fig. 9.12. Transcrystalline brittle fracture. The crack crossed the cleavage planes in the eutectic silicon precipitate and α-Al(FeMn)Si phase. In the α-aluminum solid solution the fracture was of the transcrystalline, ductile kind. Equiaxial and oval, open dimples are visible. 1200×

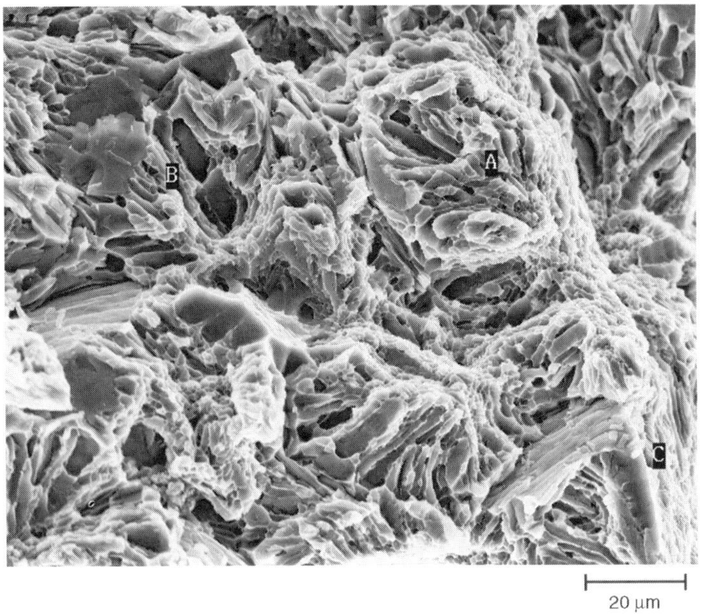

Fig. 9.19 Transcrystalline fracture of mixed morphology. The crack crossed the eutectic grains. The micronecks have formed from the deformed α-aluminum solid solution in the boundary zones (see Fig. 9.5). 750×

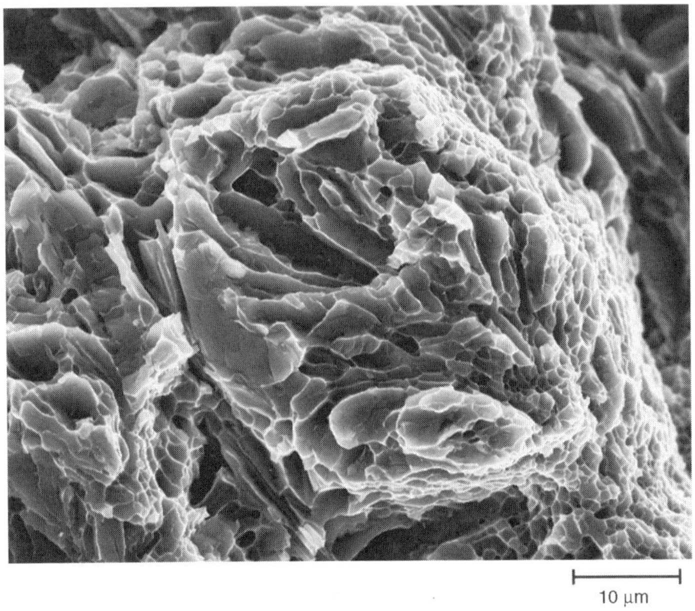

Fig. 9.20 Detail of [A] in Fig. 9.19. Fracture of mixed morphology in the eutectic grain. Cleavage facets with cleavage steps are visible in the eutectic silicon. In the α-aluminum solid solution the equiaxial dimples have formed as a result of the microdeformation process. 1700×

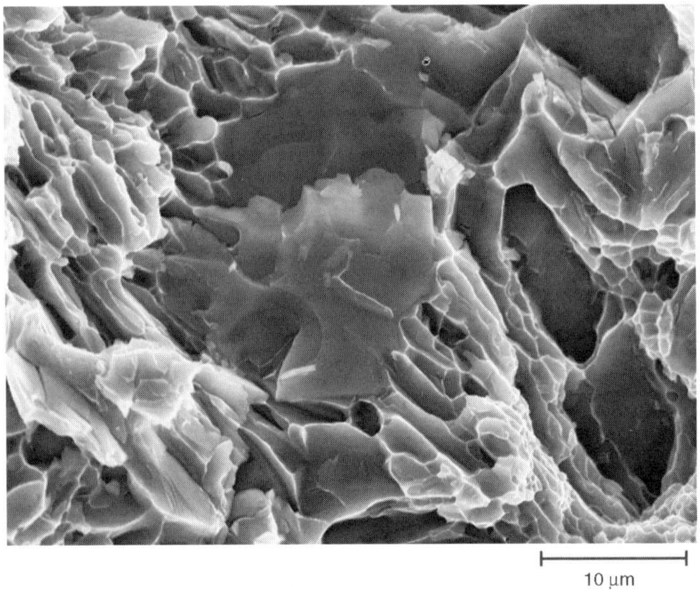

Fig. 9.21 Detail of [B] in Fig. 9.19. Transcrystalline fracture. The main crack crossed the cleavage planes in the eutectic silicon precipitate, between two zones of plastic deformation (cellular fracture). The deformed matrix has formed the cells in two-phase zone. In the silicon precipitates the steps are visible on the cleavage facets. 2100×

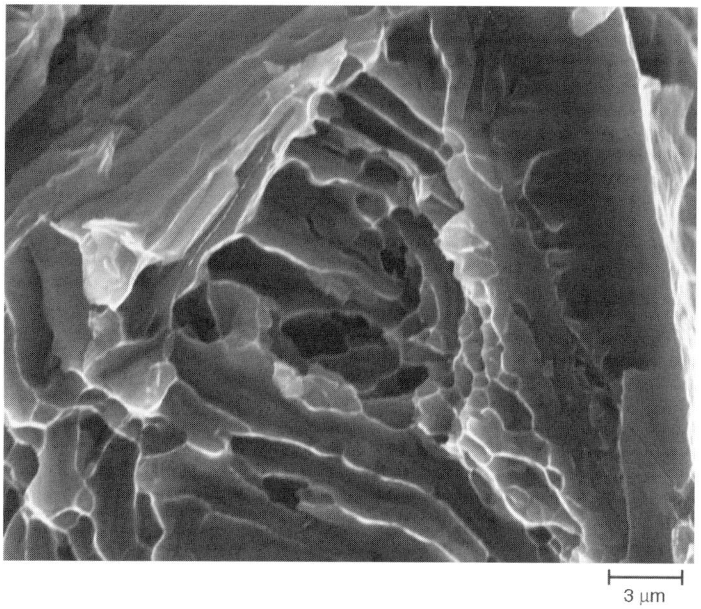

Fig. 9.22 Detail of [C] in Fig. 9.19. The cellular fracture area has formed between two micronecks in the dendrites of the α-aluminum solid solution. The visible cells are a result of its plastic deformation. 3700×

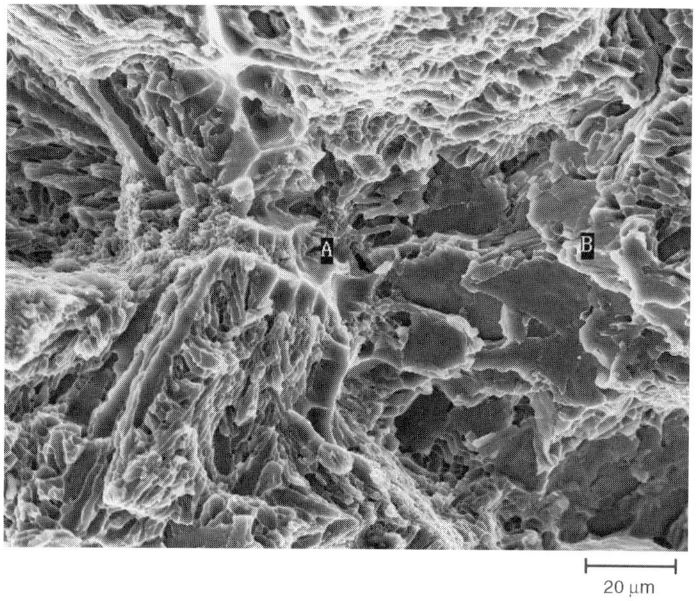

Fig. 9.23 Transcrystalline fracture of medium-developed surface. The crack crossed two-phase zone in the eutectic grain (left side of the micrograph) and the cleavage planes in the silicon precipitate (right side of micrograph). 650×

Fig. 9.24 Detail of [A] in Fig. 9.23. The fracture crosses the border zone between eutectic grains. The areas of the oval and equiaxial dimples are visible in the α-aluminum solid solution. The formation of the tear ridge in the micronecks of the α-aluminum solid solution (as a last stage of the decohesion process in this area) was preceded by coalescence of the voids and shear of the dimples. This mechanism is characteristic for ductile fracture. 1800×

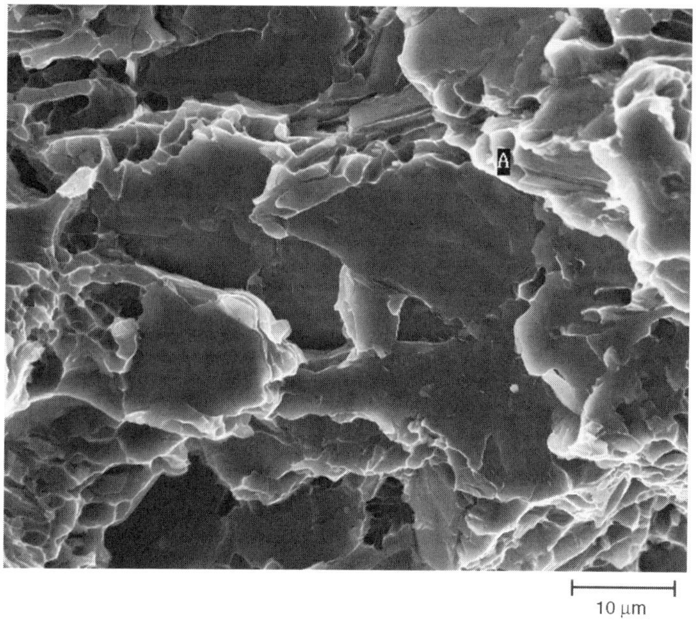

Fig. 9.25 Detail of [B] in Fig. 9.23. Transcrystalline cleavage fracture in the silicon precipitate. The crack front crossed the parallel cleavage planes, separated with the steps, forming the terraces. Traces of plastic deformation of the matrix can be observed. 1500×

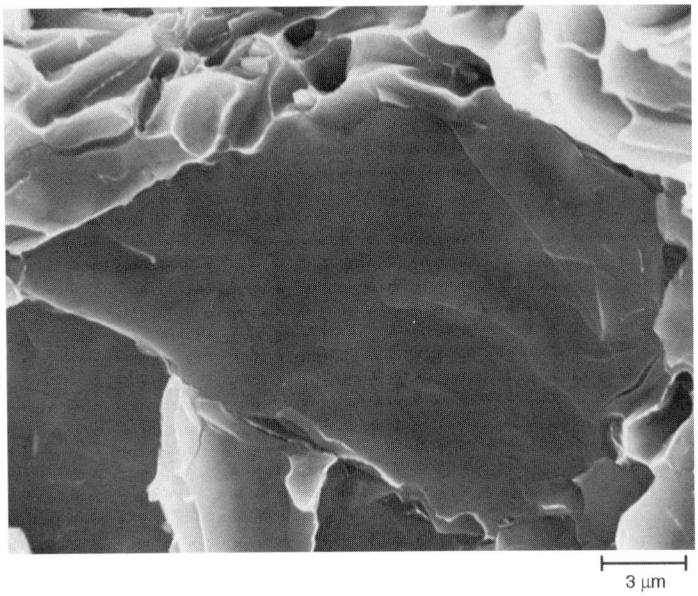

Fig. 9.26 Detail of [A] in Fig. 9.25. Transcrystalline cleavage fracture in silicon precipitate. Cleavage steps have formed the waved bands on the parallel cleavage facets. 4500×

CHAPTER 10

Material Defects on Fracture Surfaces

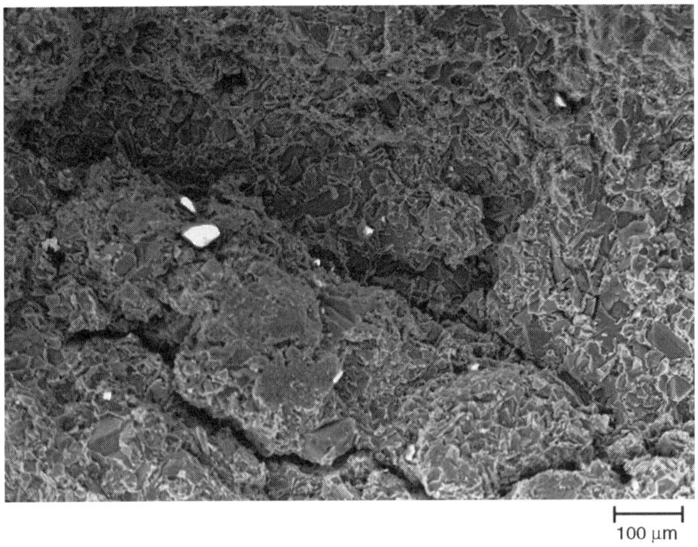

Fig. 10.1 Oxide film and the inclusion clusters on the fracture surface after static tensile test. Alloy 413.0 (AlSi11), nonmodified, permanent mold casting. 95×

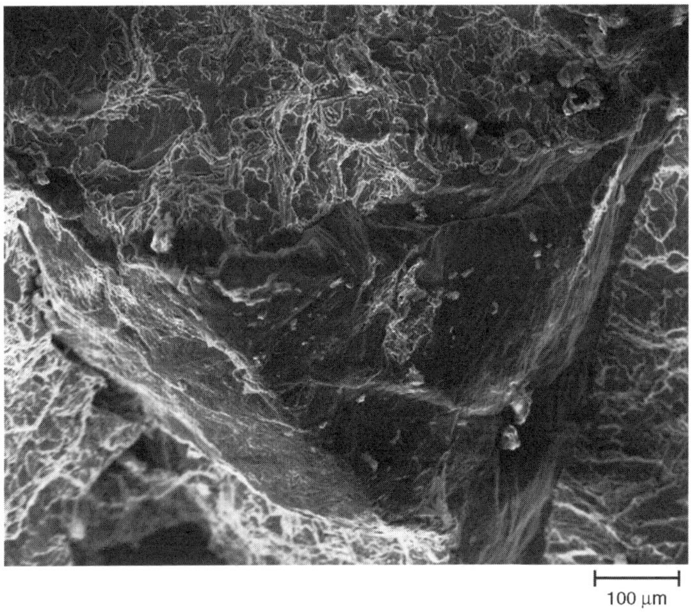

Fig. 10.2 Oxide film and cluster of the oxide inclusions near secondary crack on the fracture surface after static tensile test. The fracture in the neighborhood of the defect is transcrystalline. Alloy 413.0 (AlSi11) nonmodified, permanent mold casting. 120×

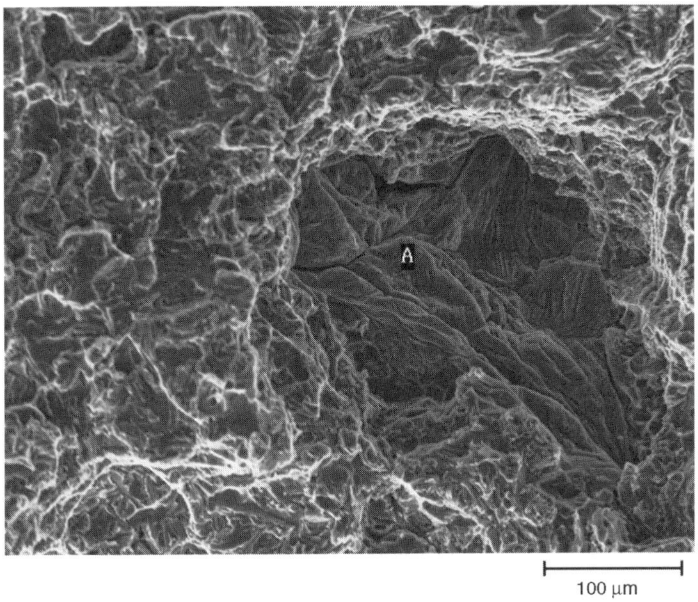

Fig. 10.3 Oxide films on the intercrystalline fracture surface after V-notch impact test at 21 °C (70 °F). Secondary cracks were formed around the defect. Alloy 356.0 (AlSi7) nonmodified, permanent mold casting. 200×

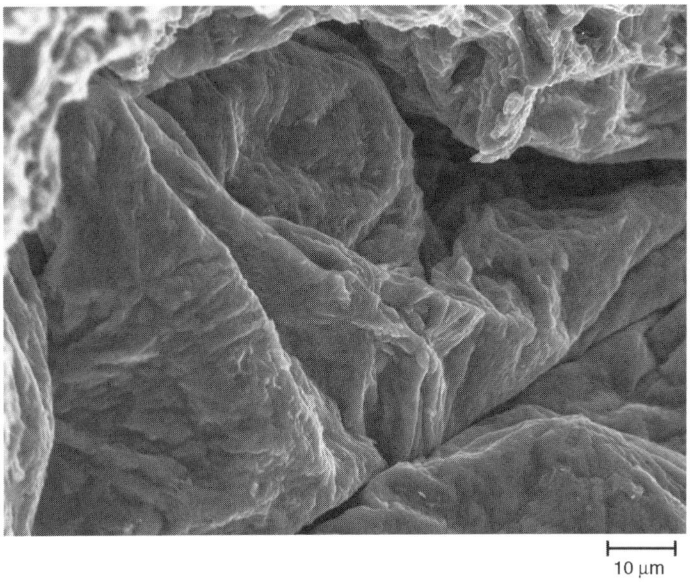

Fig. 10.4 Detail of [A] in Fig. 10.3. The intercrystalline fracture is covered with oxide film. Internal discontinuities are present on the border of the crystallites. 1000×

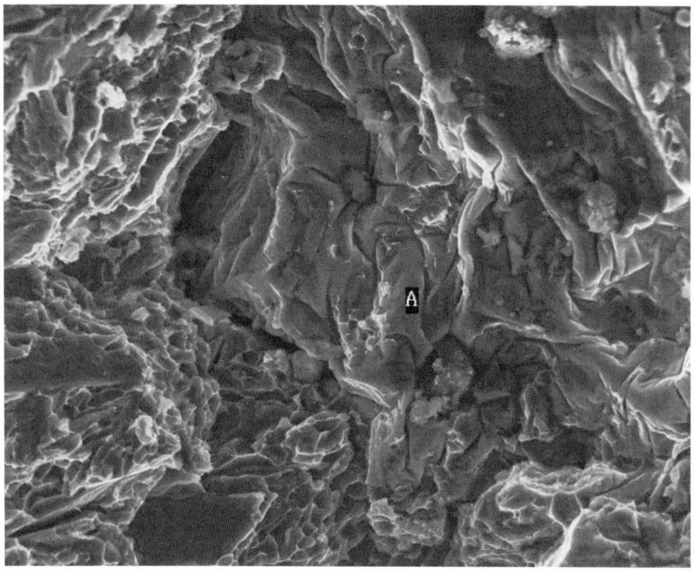

Fig. 10.5 The intercrystalline fracture is covered with oxide film (after static tensile test). The internal discontinuities are present on the borders of the crystallites. The dimples of plastic deformation are visible in neighborhood of the defect in the matrix. Cleavage facets can be observed in the silicon crystals. Alloy 356.0 (AlSi7Mg), refined, modified, permanent mold casting. 1000×

Fig. 10.6 Detail of [A] in Fig. 10.5. The intercrystalline fracture is covered with the oxide film. The internal discontinuities are present on the borders of the crystallites. The matrix was slightly deformed in the interface, in zone of the retained cohesion. 3670×

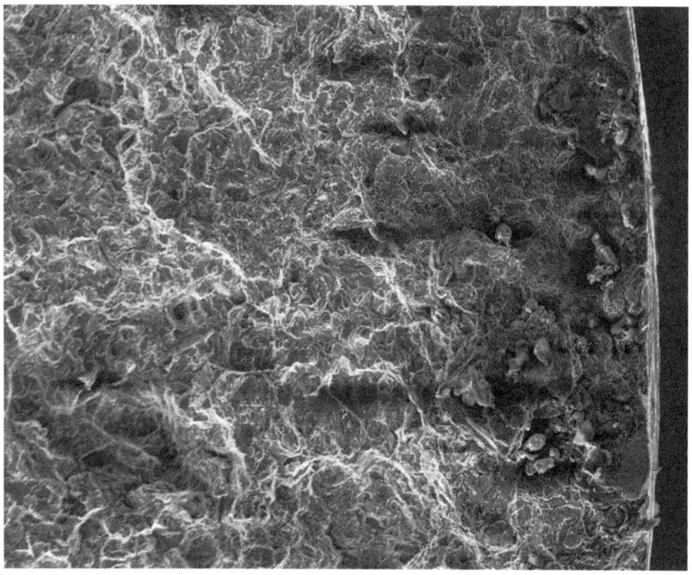

Fig. 10.7 Clusters of oxide inclusions, containing: O, Na, Mg, Al, Si, Cl, K, and Ca on the fracture surface, after static tensile test. Alloy 356.0 (AlSi7Mg), refined, modified, permanent mold casting. 100×

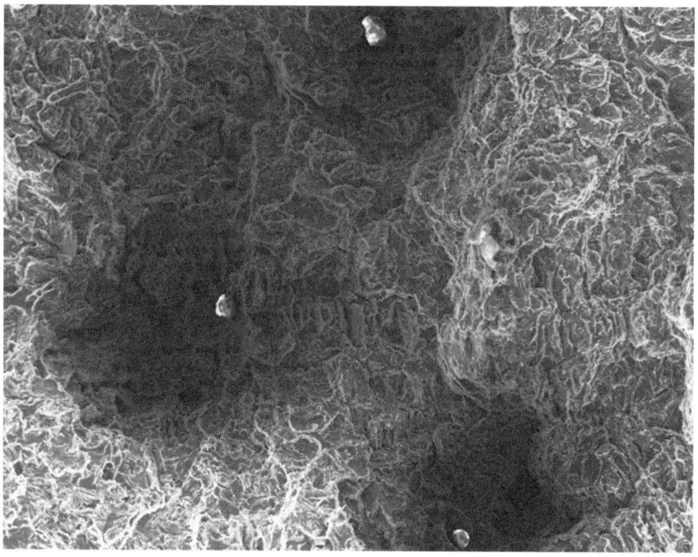

Fig. 10.8 Oxide inclusions containing N, O, Al, Si, S, Ca, and Fe situated on the bottom of internal discontinuities on the fracture surface (after static tensile test). Alloy 356.0 (AlSi7Mg), refined, modified, permanent mold casting. 100×

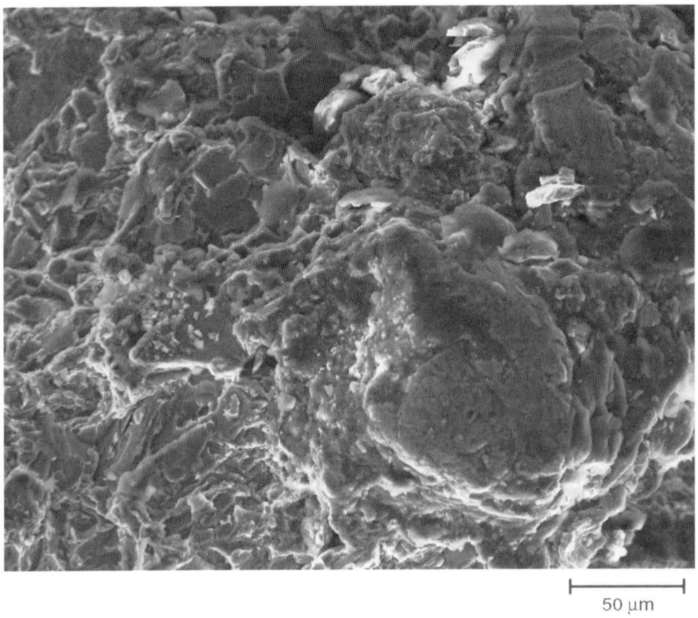

Fig. 10.9 Clusters of oxide inclusions containing: O, Al, Si, K, and Ca on the fracture surface (after static tensile test). Alloy 413.0 (AlSi11), refined, modified, permanent mold casting. 340×

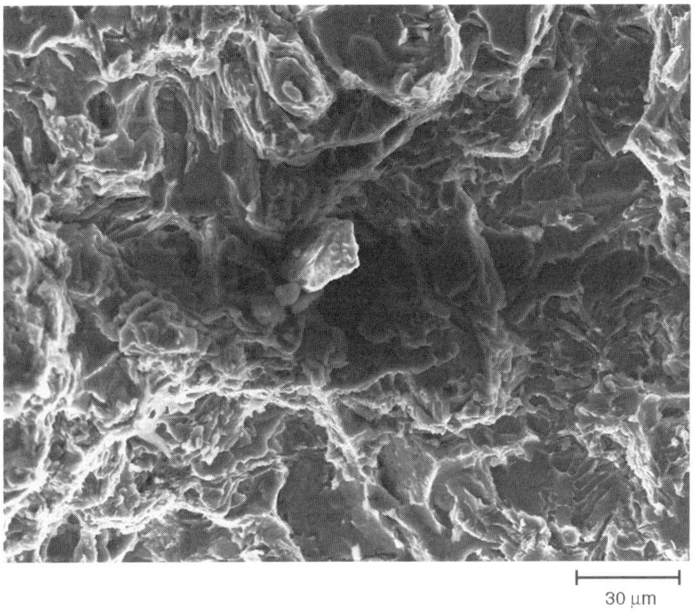

Fig. 10.10 Oxide inclusions containing O, Al, Si, P, and Fe on the fracture surface (after static tensile test). Alloy 356.0 (AlSi7Mg), refined, modified, permanent mold casting. 480×

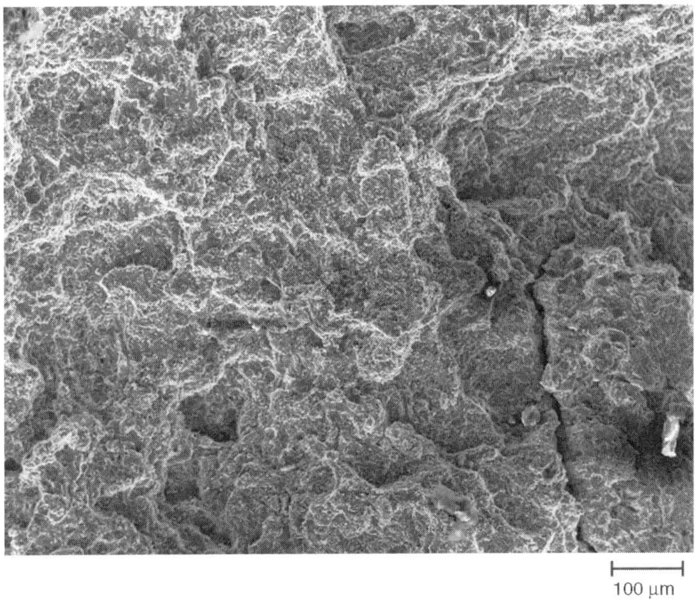

Fig. 10.11 Oxide inclusions containing Ca and Cl on the fracture surface (after static tensile test). Alloy 336.0 (AlSi13Mg1CuNi), refined, modified, permanent mold casting. 95×

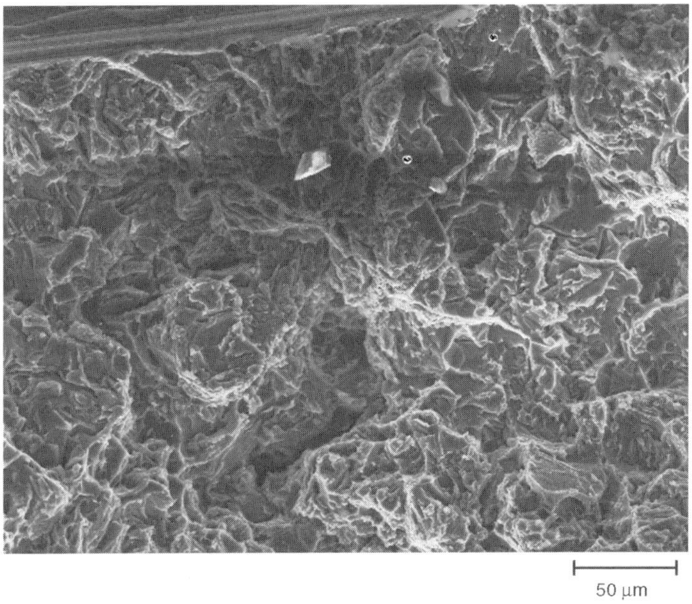

Fig. 10.12 Clusters of oxide inclusions containing N, O, Al, Si, Ca, and Fe in the interdendritic space, near the internal crack on the fracture surface (after static tensile test). Alloy 336.0 (AlSi13Mg1CuNi), refined, modified, permanent mold casting. 270×

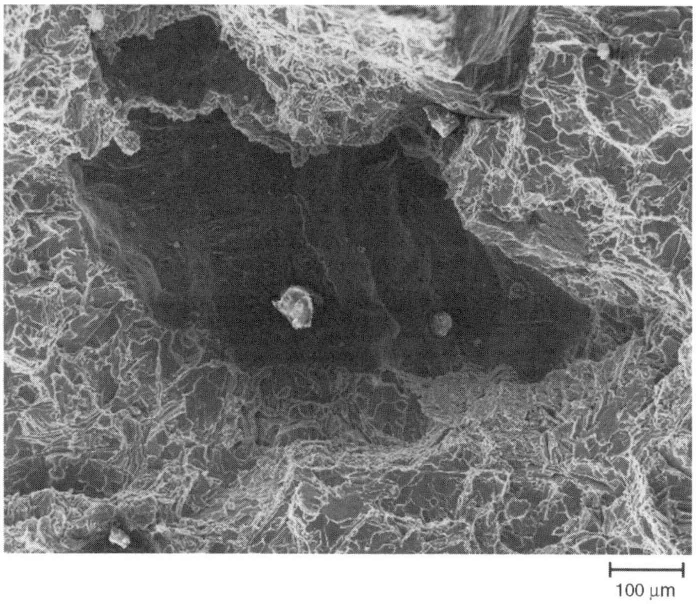

Fig. 10.13 Oxide inclusions containing O, Na, Cl, Ca, and particles of SiO_2 in the large shrinkage, covered with the oxide film on the fracture surface (after static tensile test). Alloy 413.0 (AlSi11), refined, modified, permanent mold casting. 110×

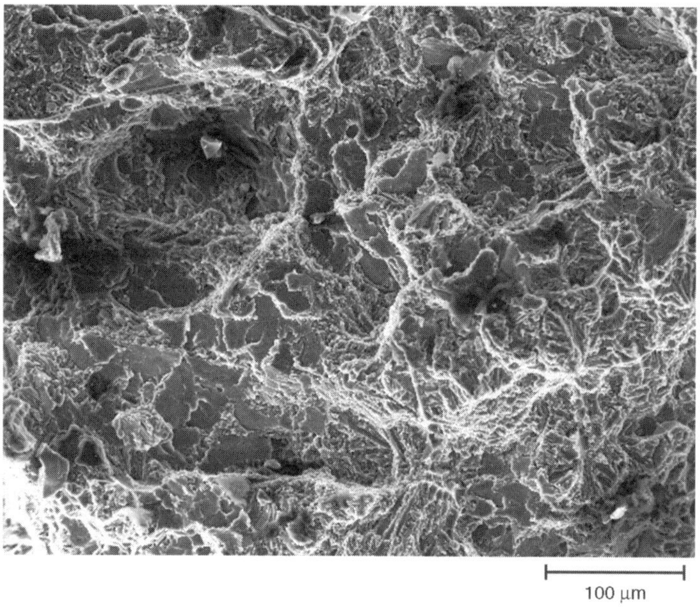

Fig. 10.14 Oxide inclusions containing Na, Cl, Ca, and particles of SiO_2 on the fracture surface (after static tensile test). Alloy 413.0 (AlSi11), refined, modified, permanent mold casting. 200×

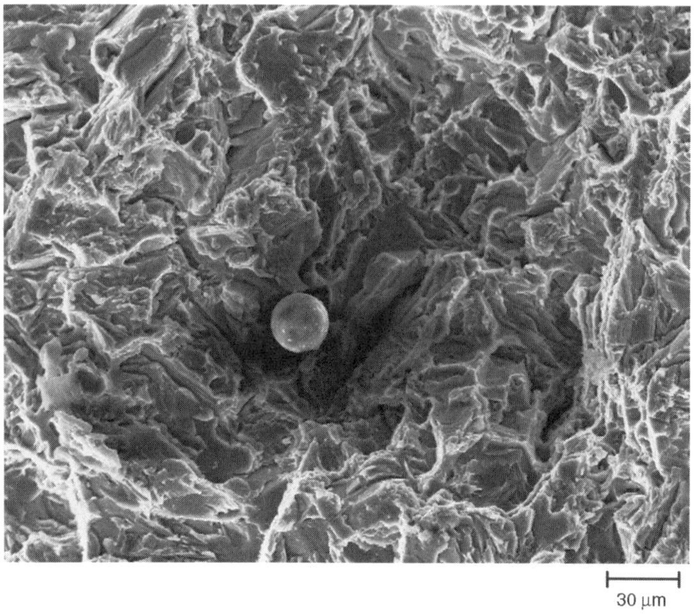

Fig. 10.15 Spherical inclusion containing O, Al, Si, K, Ti, and Fe, on the fracture surface (after static tensile test). Rectilinear secondary cracks are visible. Alloy 390.0 (AlSi21CuNi), refined, modified, permanent mold casting. 300×

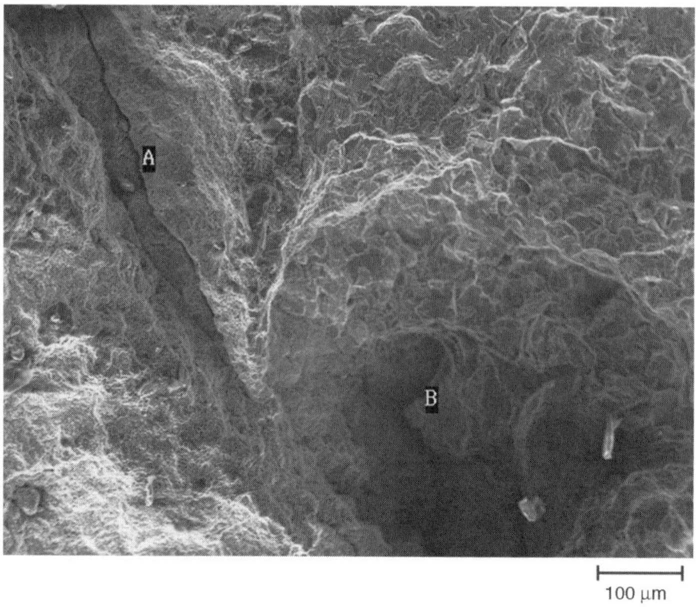

Fig. 10.16 Oxide inclusions containing Na, Cl, K, C, O, N, Al, Si, Ca, and Fe, on the fracture surface (after static tensile test). The crack formed in the zone covered with oxide film near the shrinkage. Alloy 356.0 (AlSi7Mg), refined, modified, permanent mold casting. 120×

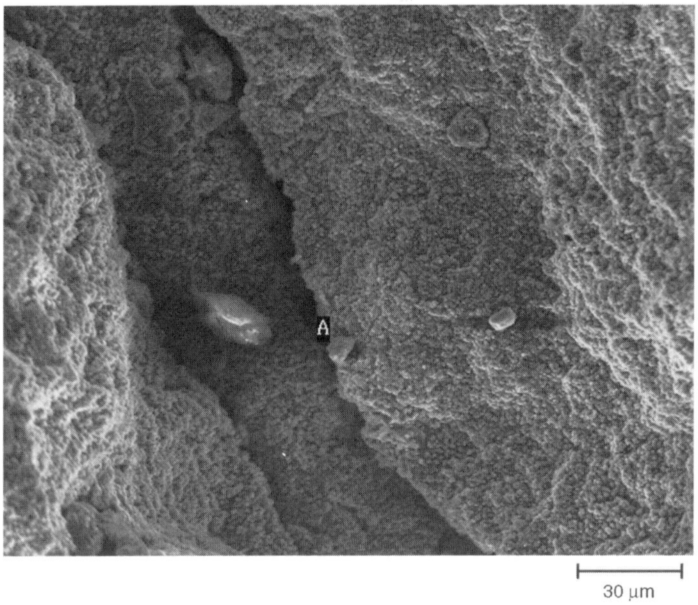

Fig. 10.17 Detail of [A] in Fig. 10.16. Oxide inclusion film covering the fracture surface. The inclusions contain O, Mg, Al, and Si (see Fig. 10.18). Particles containing Na, K, and Cl also are present in this zone. 500×

Fig. 10.18 Detail of [A] in Fig. 10.17. Morphology of the oxide inclusions O, Mg, Al, Si. 2800×

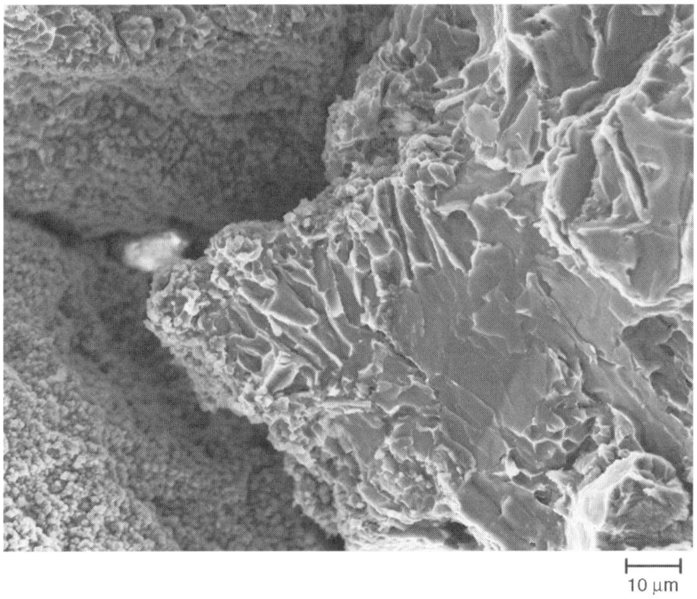

Fig. 10.19 Detail of [B] in Fig. 10.16. Internal cracks are present on the inside surface of the shrinkage discontinuity. The fracture near the defects is transcrystalline. 800×

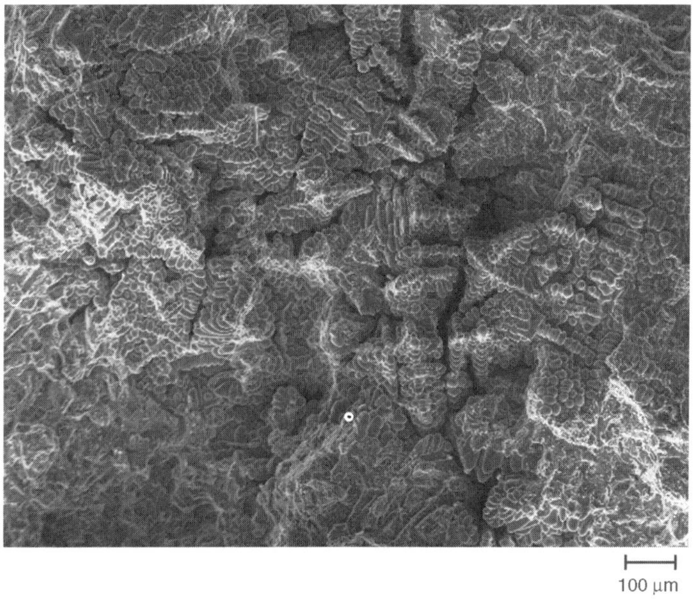

Fig. 10.20 The surface of the interdendritic shrinkage on the fracture after V-notch impact test at 21 °C (70 °F). Secondary cracks are present in the interdendritic spaces. Alloy 356.0 (AlSi7Mg), refined, modified, permanent mold casting. 70×

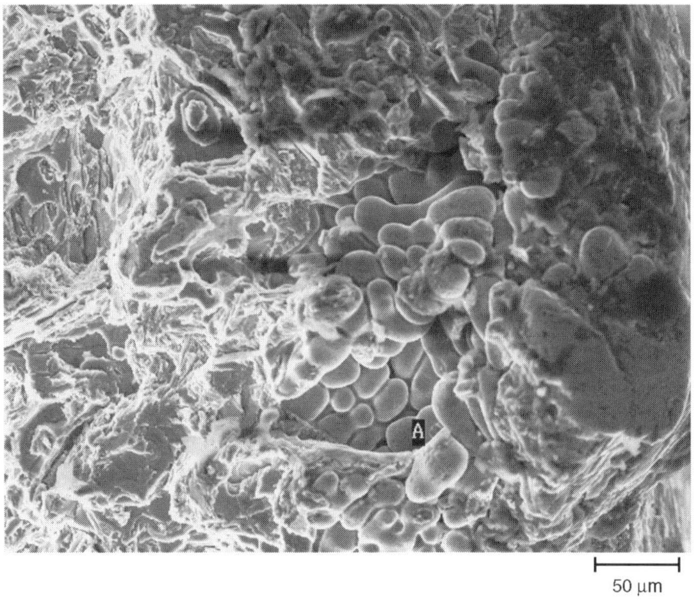

Fig. 10.21 The internal surface of the interdendritic shrinkage on the fracture after V-notch impact test at 21 °C (70 °F). Secondary cracks are present in the interdendritic spaces. Alloy 356.0 (AlSi7Mg), refined, modified, permanent mold casting. 250×

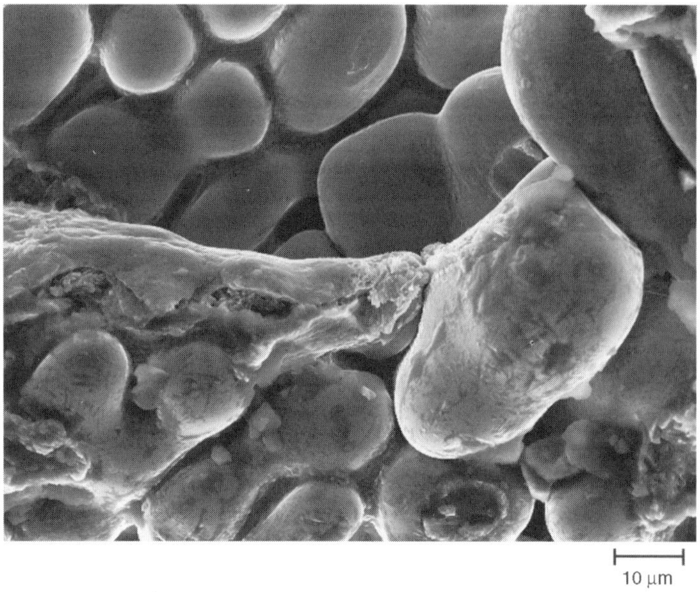

Fig. 10.22 Detail of [B] in Fig. 10.21. Morphology of the interdendritic fracture. The zones of the retained cohesion in matrix can be observed. 1000×

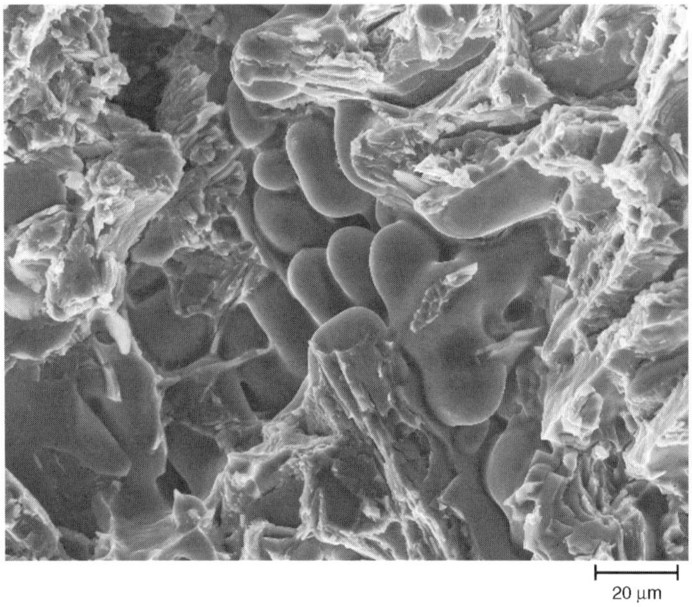

Fig. 10.23 Morphology of the interdendritic shrinkage on the surface after V-notch impact test at 21 °C (70 °F). Transcrystalline fracture is visible in the eutectic regions. Alloy 356.0 (AlSi7Mg), refined, modified, permanent mold casting. 600×

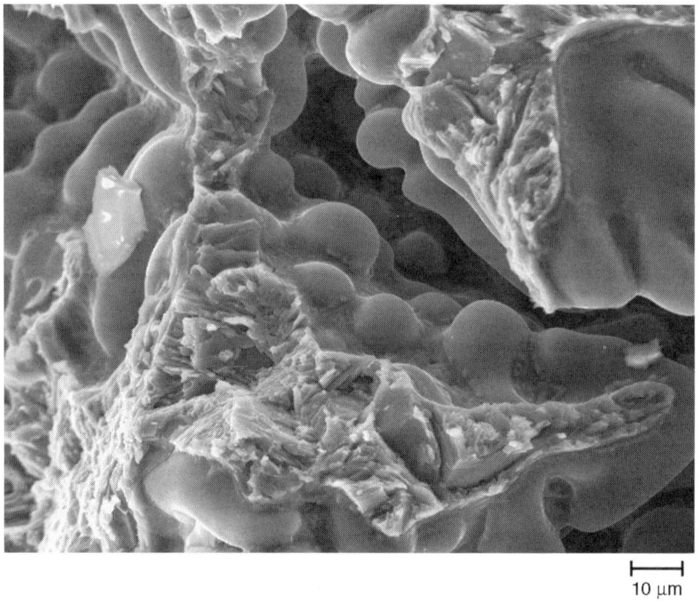

Fig. 10.24 Morphology of the interdendritic shrinkage on the fracture surface after V-notch impact test at 21 °C (70 °F). In the eutectic regions the transcrystalline fracture is visible. Alloy 356.0 (AlSi7Mg), refined, modified, permanent mold casting. 800×

Index

F

G

H

I

L

M

N

O

P

Q

R

S